BEYOND EXPECTATIONS

PRACTICAL TOOLS TO MASTER EXPECTATIONS AND BUILD STRONGER LEADERSHIP

Deirdre Morrison and Gabriella Benkő

This book is a work of nonfiction. While every effort has been made to ensure that the content is accurate, the authors and publisher assume no responsibility for errors or omissions, or for damages resulting from the use of the information contained herein.

Disclaimer: The information provided in this book is for general informational purposes only. It is not intended as professional advice and should not be relied upon as such. Readers should consult with a qualified professional regarding their specific circumstances.

Cover Illustration: Created by Deirdre Morrison using graphics from Canva. Canva content is used in accordance with Canva's licensing agreements.

ISBN: 978-1-0685396-1-9

Printed in United Kingdom

For those who expected that we would,
and those who expected that we would not.
Our gratitude to you all.

Contents

Introduction

Let's be clear from the outset:

This book isn't designed to help you rid yourself of expectations. It's designed to help you understand them and navigate them with confidence.

Expectations, like many aspects of life, are not inherently good or bad. Our goal is to help you determine whether your expectations are making you more effective or less so. Are they helpful or unhelpful in any given situation?

> "The illiterate of the twenty-first century will not be those who cannot read and write, but those who cannot learn, unlearn, and relearn."
> — Alvin Toffler

Our brains are capable of profound change—thanks to **neuroplasticity**. This is the term used to describe how our brains continue to adapt and reshape throughout life. In relation to expectations, we can think about it this way - our brains learn certain things, and we are conscious of some, and not conscious of others. The same applies to our expectations. Expectations become part of our thought processes, and the more we think those thoughts, the more our brain learns them - that's why we sometimes struggle to shake off certain habits or ways of thinking.

The good news? Once you start to see how the brain operates, it opens up a world of possibilities. You can choose better actions, reactions, and interactions by tapping into new information and thought processes. This creates space for growth in your leadership and personal development—step by step, mastering your expectations.

Why You Picked Up This Book

You're likely reading this because you are ambitious, intelligent, and driven to create positive change—whether in your professional life, your relationships, or within your organisation.

This book is designed for those who want to *lead* that change. Whether it's within a business you're building, an organisation you're part of, or another group setting, these insights and tools will help you excel.
Even if you've been honing your leadership skills for years, there's always room to elevate your approach. That's why we've packed this book with fresh perspectives, giving you tools and frameworks to build stronger leadership through an applied neuroscience lens.

How This Book is Structured

We won't waste a moment of your valuable time. We're not about fluff. This book is concise, practical, and focused on preparing you to use tools that will help you quickly level up your leadership skills. It's designed to be experiential, so you can dive into exercises and examples that bring the theory to life.

Something else you should know - this is not a 'science book', although it is steeped in applied neuroscience, and drawn from years of observation and practise. We use 'Bite-Size Brain

Science'—key concepts delivered in a digestible format—paired with hands-on exercises so you can test and apply the ideas for yourself. These aren't just theories; they're actionable steps that will contribute to your leadership mastery over time.

In the real world, leadership is built on cumulative moments, small shifts in understanding that drive greater impact.

Reflection Point: As you move through the book, consider:

How are your current expectations shaping your actions?

What small, actionable steps can you take to master them and grow in your leadership?

What You'll Gain

By the time you've finished this book, you will have tools and frameworks to:

- ☐ **Identify and track** expectations that arise in different situations.
- ☐ **Distinguish** between helpful and unhelpful expectations.
- ☐ **Make active, conscious choices** about how you manage and adapt expectations in real time.
- ☐ **Refine** your leadership style using applied neuroscience insights.

Leadership, like mastering expectations, is a journey. This book is a part of that process, giving you both immediate tools and the knowledge to keep evolving. We'll also point you toward additional resources for continued growth and support, because mastery doesn't stop when the book ends.

REFLECTION POINT:

☐ **Think about the expectations you currently have for yourself as a leader. How do they align with where you are now, and where you want to go?**

The Path to Mastery

Expectations mastery is not a one-time fix. It's a skill that develops over time with consistent reflection, adjustment, and practice. Supporting our neuroplasticity as we learn and adapt, requires patience, self-care and self-compassion. It is incredibly unlikely that you will consistently apply everything you read here immediately and consistently. But commitment to and focus on your growth will pay off. Every moment of leadership offers an opportunity to apply and refine the techniques you'll learn in this book. Value them all as the learning opportunities they are.

To summarise, by engaging with this book, you will:

☐ Experiment with hands-on techniques to **master expectations** and apply them to leadership.
☐ Learn neuroscience-based insights in an **accessible** format.
☐ Explore **real-world case studies** that ground theory in practice.
☐ Reflect on your leadership actions, reactions, and interactions to foster continuous growth.

What Not to Expect

You are a smart and capable individual. You already know that there are no magic wands, and this book isn't a quick fix. But it is a quick start. The tools are simple but require ongoing use and reflection to achieve real change. And while we cover a lot of ground, we can only paint a broad picture with sample scenarios in the confines of this book. But we are here to support your leadership journey beyond these pages. Whether through additional resources, 1-1 development work, or further training, feel free to reach out.

Lastly, this book is not a substitute for professional therapy. If you are dealing with trauma or its effects, please seek support from a licensed professional.

REFLECTION POINT:

- ☐ **Consider your journey through this book as a start to mastering your leadership.**
- ☐ **What are your expectations about the pace of your growth?**
- ☐ **How can you set yourself up for long-term mastery?**

"The road to nowhere is paved with a-ha moments.
The road to progress is paved with a-ha actions."

A Journey Built on A-Ha Actions A-ha moments are great. They

can be excellent catalysts for
change. But without application, they do not guarantee any
change or improvement. **The road to mastery is paved with small,**

consistent actions. The
a-ha moments you experience are just the beginning—the real
progress comes when those insights turn into actionable steps.

Remember, leadership mastery isn't a final destination; it's a
continuous process. Each step forward enhances your capacity
to manage expectations, lead effectively, and drive impact.

*Note: The case studies in this book have been amalgamated
from real-world situations, without compromising the integrity of
the scenarios or outcomes. Names and details have been
changed to protect privacy.*

Deirdre Morrison and Gabriella Benkő

Expectations in Real Life: Lessons from Personal Experience

DEIRDRE'S EXPECTATIONS: Scaling the Heights

Many years ago, while on a trekking trip in the foothills of the Himalayas in Nepal, I realised that expectations were shaping how my experience of this once in a lifetime trip.

Coming from a pretty flat country (Ireland's highest peak is Carrauntoohil, at 1038 metres above sea level), I had no experience of really mountainous terrain. In comparison to the towering Himalayas, 1038m looked like a speed bump. After a difficult first day of seemingly endless climbing, my understanding of trekking was challenged, to say the least!
Because I was unprepared for this very literal 'levelling up', the first day was a struggle. I had expected a vigorous walk, and some dramatic scenery. What I got was a test of stamina and fitness.
I asked Raj, the guide, if the next day would be more of the same. "Tomorrow, almost gradual," he smiled warmly. Raj smiled a lot. This did not seem like a satisfactory answer. I did not know what to expect from 'almost gradual'. I rephrased my question: "Will tomorrow be harder or easier?" "Bit harder," he replied with that same genuine smile.
That was all I needed to know. Now I could set my expectations for the day. I knew what I'd done so far, and I knew I'd need to do the same and a bit more the next day, so I adjusted. Despite the incredibly beautiful scenery all around me, I revised my expectations of a gentle sightseeing stroll, stuck a bunch of Buddhist chants on my headphones, and got stuck in.

Interestingly, Day 2 ended up being a bit easier, not harder—at least in my memory of it. That's because my expectations were aligned with the reality of the Himalayas, not my uninformed imagination based on Ireland's modest hillocks.

This experience illustrates a powerful analogy for many things—including leadership. When we're not fully aware of the effort, energy, and approach required, our misaligned expectations can sideswipe us, knock the wind out of us, and rob us of our confidence. Recalibrating expectations helps us regain perspective, budget our energy, and form a realistic view of the task at hand. When we master the art of setting realistic expectations, we lay the foundation for stronger leadership and more mindful decision-making.

REFLECTION:

- ☐ **When has a situation derailed you because your expectations—verbalised or otherwise—created an imagined version that didn't reflect reality?**
- ☐ **What was your response?**
- ☐ **Would you have done something differently if you had recognised and realigned your expectations sooner?**

GABRIELLA - Aperitif Expectations

When I moved to France, I was eager to immerse myself in the culture and make new friends. So, when a neighbour from my building invited me to an *apéro*—a casual evening gathering with drinks and snacks—I was thrilled. The invitation said 7 p.m., and

I wanted to make a good impression. Coming from a Hungarian family where punctuality was not just a virtue but an obligation, I knew exactly what to do: arrive at 7 p.m. sharp.

I rang the doorbell and waited. A minute passed, then two, and soon five minutes had gone by. As I stood there, I started to feel a bit uneasy. Did I get the time wrong? Was I at the wrong door? Just as I was about to check my phone again, the door opened. My host stood there, clearly surprised, with wet hair and a bathrobe loosely tied around her. She had just come out of the shower. I could see the mild shock on her face, though she quickly masked it with a friendly smile.

"Ah, bonsoir!" she said, her voice carrying a mix of warmth and confusion. "You're here early!"

I stammered an apology, explaining that I thought 7 p.m. meant, well, 7 p.m. She laughed softly, ushering me in and explaining that in France, arriving on time to an *apéro* is considered early. Most people come about 20 minutes later. Feeling a bit embarrassed, I sat down, surrounded by neatly arranged snacks and untouched drinks, and waited. We made small talk, but I could sense her slight discomfort, perhaps wishing she had a few more minutes to prepare.

As the clock ticked towards 7:15 p.m., the first of the other guests started to arrive. By 7:20 p.m., the apartment was buzzing with conversation and laughter. I watched as people trickled in, casually and unhurriedly, greeting the host with kisses on both cheeks. I realised that my early arrival was more of an anomaly than the norm.

Over time, I learned that in France, being "on time" for social gatherings is a flexible concept. Arriving exactly at the designated hour, especially for something informal like an *apéro*, is often seen as rigid. Now, when I'm invited to an *apéro*, I make sure to arrive about 20 minutes later, blending in with the flow of other guests.

But when I go back to Hungary and join my family's gatherings, I'm always spot-on time, just like I was taught.
It's fascinating how the same concept—being polite—can have different interpretations, depending on where you are.

Expectations vary with culture, teaching me that politeness is about understanding and adapting to different customs.
Leadership is like understanding cultural expectations at a gathering. It requires both an awareness of others' norms and self-awareness of your own actions. Just as arriving at the right time shows you understand the host's expectations, effective leadership involves recognising the needs of your team while being mindful of how your actions and decisions impact them. Mastering the art of adapting expectations to fit the context can be the key to stronger, more effective leadership.

REFLECTION:

☐ **When have your expectations of social norms led to an uncomfortable or surprising situation?**

☐ **How did you adjust when your initial understanding of the situation didn't match the cultural reality?**

☐ **What can you do to better align your expectations with the cultural or team norms in new environments?**

PART 1

Getting to Know Expectations: The First Step Toward Mastery

Let's Begin...

If you're a living, breathing human, you have expectations. And so do the other living, breathing humans around you. Whether you notice them or not, expectations colour much of our life experience. How well we identify, manage, and ultimately master them shapes our levels of satisfaction and fulfilment, both personally and professionally.

In leadership, **mastering expectations** is crucial. It is not just about being aware of them but knowing how to effectively manage them or interrupt them to enhance decision-making, communication, and overall success.

'Managing expectations' is a phrase often used, especially in business and leadership. But what exactly are we managing? Can we really manage something we don't fully understand? And how do we move beyond simply 'managing' them?

Expectations, assumptions, and predictions are closely related phenomena, but they are also distinct enough to warrant individual attention. Understanding these distinctions is the first step in the journey toward mastering expectations.

- ☐ **Expectation**: A strong belief that something will happen or be the case.

- ☐ **Assumption**: A thing accepted as true or certain to happen, without proof or evidence.

- ☐ **Prediction**: A statement about what you think will happen.

Reflecting on these distinctions can give you a deeper insight into how they affect your behaviour and interactions.

REFLECTION EXERCISE:

- ☐ **Notice the differences in these descriptions.**
- ☐ **Have you ever thought about these distinctions before?**
- ☐ **How do they relate to your daily experiences?**
- ☐ **What is it about these terms that strikes you?**

Take a moment to write down your insights and impressions, as they may reveal key aspects of how you interact with others and currently manage expectations in your life.

We rarely stop to consider the roots of our expectations, and it's even more unusual to have tools to diagnose and adjust the impact of unmet expectations. Yet, mastery over our expectations comes when we learn to **anticipate and diffuse** them before they negatively affect our performance, relationships, and satisfaction.

It is often the case that expectations are only noticed when they're unmet. We experience disappointment, frustration, or even anger. However, expectations can sometimes be exceeded, and we are surprised or delighted. Throughout this book, you'll be invited to explore which outcomes dominate your focus—and how that impacts your growth as a leader.

> "The problem is not the problem,
> the problem is your attitude about the problem."
> *Captain Jack Sparrow, Pirates of the Carribean*

Mastering the Inner Dialogue of Expectations

As Captain Jack Sparrow so wisely observed, it is very often in our perspective and approach to a problem that much of the 'problem' exists. In much the same way, expectations in themselves are not the issue—it's how we **respond to them**.

Expectations aren't inherently good or bad. Like so many of our brain's processes, they are neutral. We judge them to be good or bad, depending on context, circumstance, and other factors. The thoughts, feelings, and emotions that surround our expectations are driven by brain processes designed to keep us alive and safe. When we understand these processes, we can more easily shift our mindset, allowing us to navigate challenging situations with greater clarity and effectiveness.

We'll delve deeper into these processes later in the book. For now, remember that expectations can be **helpful or unhelpful**, **productive or disruptive**. Mastering expectations is about learning to **optimise your leadership and performance** by recognising which expectations are driving you toward success and which are pulling you away from it.

Mastery Through Communication

Another key to mastering expectations lies in our ability to communicate them. Mastery in leadership requires clarity in how we convey our needs and wants. When there are **gaps** between our expectations and others' actions, it can lead to frustration. As you continue your journey toward mastery, it's vital to recognise that closing these gaps - through clear communication - directly reduced the risk of unmet expectations.

REFLECTION EXERCISE:

- ☐ **How do you currently communicate your expectations to others?**
- ☐ **What do you fail to communicate that leaves others uncertain or unaware of what you expect?**
- ☐ **How can you improve your communication to mitigate gaps between expectations and actions?**

"The single biggest problem in communication is the illusion that it has taken place."
George Bernard Shaw

Deirdre Morrison and Gabriella Benkő

REFLECTION POINTS ON MASTERY:

- ☐ **How has your understanding of expectations shifted in this section?**
- ☐ **What new insights have you gained about managing your own expectations and communicating them more clearly?**

By developing these foundational skills, you'll take the first step on the path to mastering expectations—allowing you to lead with clarity, resilience, and confidence. Each reflection and exercise helps build the tools necessary for mastering not just expectations, but your own growth as a leader.

CASE STUDY: When Politeness Becomes Miscommunication

Let's look at a scenario where asking for something didn't quite go as planned.

Mark manages a team of 14. He is known for his gentle manner and cares deeply about his direct reports. Most of Mark's team share his cultural background, so their communication styles are familiar—they speak similarly and tend to follow the same social cues of what's considered polite or impolite.

However, a new team member recently joined from a different country. She is excellent at her job but communicates differently from Mark and the rest of the team. One day, Mark was dividing tasks on a project. After assigning work to the other team members, he turned to the new colleague and said, "Would you like to finalise the details for the report?"

To everyone else in the room, this was Mark's polite way of instructing her to complete the task. But the new team member, interpreting the question at face value, responded, "No, I will be busy finishing another task."

Mark was taken aback. Eyebrows raised all around the room. He had asked nicely, so what went wrong?
This is a classic case of an assumption getting in the way of expectations being met. Mark assumed his polite request would be understood as an instruction. However, the new team member's cultural background led her to interpret it as an actual question—a request that she could decline.

It's an easy mistake to make. But understanding these misalignments allows us to recalibrate our communication, recognising that expectations are often **unspoken assumptions** that differ based on culture, context, or individual experiences.

Managing Expectations in Action

In this instance, Mark's expectations were shaped by his 'normal'—the communication style he was accustomed to. However, there were bigger expectations at play—cultural expectations that were unseen but powerful.

Let's look a little closer at Mark's situation.

Initially, he might have felt undermined. After all, his new colleague declined to take on a task he required and expected her to do. At that moment, Mark's brain was likely assessing whether this posed a **threat** to his authority or status. Our brains take the mission of protecting us very seriously, and when expectations are unmet, it can feel scary, or threatening, activating danger signals in our brains.

If Mark had unconsciously decided that his status was threatened, he might have reacted defensively by asserting authority, or perhaps even ignoring the situation. Either response could have made things worse. However, through personal development, Mark had cultivated the **presence of mind** to recognise that his expectations might not apply universally, because they were shaped by his own norms.

This allowed him to pause and choose a more effective response. Instead of reacting out of frustration, he congratulated his colleague for being productive early in the project and recognised her workload. Mark used this opportunity to foster teamwork, asking another colleague to assist with both tasks and later, scheduling a conversation to explore their communication styles.

Mark's ability to recalibrate expectations not only helped him avoid conflict but also enabled him to **strengthen team dynamics**. He communicated his expectations clearly, equipping his colleagues to succeed together in a supportive environment.

Understanding Our Brain's Role in Expectations

When our expectations aren't met, our brain can signal a threat. Its core function is to keep us safe—both physically and psychologically—and it takes this mission seriously. Unfortunately, when we operate from a **threat or fear-based state of mind**, we limit our access to higher-level thinking.

We focus on immediate concerns, losing sight of a broader, more strategic perspective.

This is why learning to manage expectations is so critical. When we cultivate the presence of mind to pause, assess, and choose a thoughtful response, we access our brain's higher-order thinking. This allows us to be more effective in our actions, reactions, and interactions.

Fear, disappointment, rejection, frustration, and anger are all common responses when things don't go as expected. But as we explore in later sections, applied neuroscience tools can help us navigate these emotions. By enhancing our awareness and presence, we develop the ability to master our responses and lead with clarity and resilience.

Section Summary: Mastering the Art of Managing Expectations

In this section, we explored the foundational concept of expectations and how they influence both personal satisfaction and professional performance. By gaining a clearer understanding clearer understanding of expectations, we lay the groundwork for developing the skills necessary to manage them — an essential aspect of leadership.

Key Concepts:
Expectations, Assumptions, and Predictions:

- ☐ **Expectations:** A strong belief that something will happen, which often guides our actions and emotional responses.

- ☐ **Assumptions:** Something accepted as true without concrete evidence, often leading to misalignment.

- ☐ **Predictions:** Statements based on what we think will happen, shaped by past experience or data.

- ☐ These related concepts influence our decision-making and interactions with others. Recognising the differences helps us approach situations with greater clarity and awareness.

Managing Expectations:

- ☐ Managing expectations is a vital skill in leadership and life. But to master this, we must first understand what expectations are, where they come from, and how they differ from assumptions and predictions.

☐ Often, we only notice our expectations when they go unmet, leading to feelings of disappointment or frustration. Building awareness of our expectations helps us anticipate potential mismatches and address them proactively.

The Role of Communication:

☐ Misaligned expectations frequently result from poor oammunication. **Consciously articulating expectations** to others can prevent misunderstandings, ensuring that assumptions don't go unchecked.

☐ As seen in Mark's case study, cultural differences and indirect communication styles can lead to significant gaps in expectation management. Recognising these factors helps leaders navigate more effectively.

The Brain's Response to Unmet Expectations:

☐ When expectations are unmet, our brain often perceives this as a threat, activating a survival response. This can limit our ability to engage in **higher-level, strategic thinking**.

☐ By understanding the brain's perception of and response to threat, we can develop greater emotional awareness and manage our reactions more thoughtfully. . This is essential for mastering emotional intelligence and fostering resilience in leadership.

Tools for Managing Expectations:

- ☐ Often overlooked tools,, such as **reflection**, are designed to help you pause, reflect, and recalibrate your expectations in real-time.

- ☐ By mastering these tools, you will develop the ability to communicate your expectations effectively, ensuring they are aligned with reality and the people you work with.

- ☐ Reflection is a foundational tool for working with the more specific tools that we provide later in the book.

REFLECTION POINTS FOR MASTERY:

- ☐ **Is your communication effectively communicating your expectations to others?**

- ☐ **Are there any assumptions influencing your communication or expectations that could lead to misunderstandings?**

- ☐ **What steps can you take to clarify them?**

Mastery Insight:

This section marks the beginning of your journey to mastering expectations in leadership and life. By gradually integrating these concepts and tools into your daily practices, you'll develop a stronger foundation for effective communication, decision-making, and emotional intelligence. With each reflection, you are building the skills necessary for long-term growth and leadership success.

PART 2
Sources of Expectations

Here's where we get into the main event —what you've been waiting for - the sources of expectations.

It feels satisfying when we finally arrive at what we've anticipated, and our expectations are being met, doesn't it?

In this section, we're going to **break down four key sources of expectations** and start identifying how they look, sound, and feel in our everyday lives. Later, you'll discover how to **evaluate these expectations**, determining whether they serve you or hold you back.

We'll also explore the idea that expectations—regardless of their source—can be either helpful or unhelpful. Developing mastery over expectations starts with understanding their roots and **recognising their impact** on your actions and decisions.

Where Do Expectations Come From?

There are many sources of expectations, arising from both internal and external sources. In some cases, we will find that we have internalised expectations that originally came from external sources. In this section, we'll untangle some of those expectations and their origins. We encourage you to notice how expectations from these various sources show up in your own life and leadership journey.

1. Expectations We Have of Ourselves

We all carry expectations for ourselves, and they can be both helpful and unhelpful. It's in this personal space where we face the most direct impact of these mental frameworks.

They often show up as the '**shoulds**' in our lives. For example:

- ☐ I should go to my colleague's leaving party.
- ☐ I should exercise more.
- ☐ I should get a promotion.
- ☐ I should be more successful by now.
- ☐ I should be more organised.

Notice how the word 'should' suggests a **hesitation**—a task or standard that we might be avoiding, not because it isn't achievable, but because there's resistance on some level. Sometimes the source of this resistance requires deep reflection and probing to uncover. Often, we discover that this resistance is rooted in an expectation that may not align with our current priorities or reality.

Just because something falls into the 'should' category doesn't make it automatically **helpful** or **unhelpful**. It's the context and the way we approach these expectations that matter.

- ☐ **Helpful:** An expectation that "I should exercise more" could keep you physically fit and contribute to better brain function and energy.

- ☐ **Unhelpful:** An expectation that "I should be more successful in my career by now" could lead to undue stress and an unsustainable workload.

REFLECTION EXERCISE:

- ☐ **What are your common "shoulds"?**
- ☐ **How do you feel about each of them - which emotions do they bring up?**
- ☐ **Consider how they influence your day-to-day actions.**

In contrast to the 'I shoulds', there are **expectations** we place on ourselves that are constructive and empowering. These expectations help us develop, learn, and grow.They can **drive us forward**, enhancing our self-efficacy and reinforcing our ability to meet challenges. Often, we frame them as beliefs, or self-beliefs.

- ☐ *I expect / believe I will be able to adapt to the new system.*
- ☐ *I expect / believe I will make thoughtful, well-considered decisions.*
- ☐ *I expect / believe I will enjoy working with and learning from my new boss.*

REFLECTION:

- ☐ **What helpful expectations do you have of yourself?**
- ☐ **How do these expectations support your growth?**

☐ **Does your inner dialogue refer to these as expectations, beliefs, or shoulds?**

2. Expectations Rooted in Society

Beyond our personal expectations, there are those embedded in the fabric of society—**unspoken agreements** about how we should behave and how we conduct our relationships with others.

For example, there's a common expectation that we will **care for or help others**, and in turn, that others will care for or help us. How far this extends beyond our closest relationships can vary greatly across both time and culture, but these **social expectations** are deeply rooted. Another typical expectation is that we will be useful in some way. The extent to which we are aware of these expectations or experience them as autonomic responses could fill an entirely separate book!

Ages and Stages: Expectations Over Time

Some of our expectations are linked to our **life stage**. As children, we develop expectations about what we should (or should not) be able to do, and these early patterns often persist into adulthood. A toddler, for example, may expect to be able to do things their parents or older siblings do, believing they 'should' be capable—imitation, trial, and error is part of the learning process.

As we grow, our expectations evolve. A five-year-old who insists on doing things **'all by myself'** is motivated by both a desire for independence and the recognition that their efforts are rewarded.

At any age, **praise** and **recognition** can boost 'feel good' neurotransmitters in our brain, reinforcing behaviour and learning.

For some, this **need for recognition** becomes a habit. As adults, we may still feel rewarded when we do things independently. However, the flip side of this is that it can make asking for help difficult, leading to unrealistic self-expectations. The idea of asking for assistance may feel like a sign of weakness or vulnerability because we've learned to expect that we can - or should- handle things alone.

Conversely, as we grow older, we may struggle with **outdated expectations** of ourselves.

For instance:

- ☐ Our physical speed or strength may have diminished.
- ☐ Family responsibilities may limit the time we have to devote to our career.
- ☐ We may need glasses or other aids as our senses decline.
- ☐ Financial shifts may mean we can't enjoy the same lifestyle as before.

Outdated expectations can lead to frustration or self-criticism if we don't pause to **reassess** their relevance. Instead, we need to adopt a practice of **self-compassion**, recognising that not all expectations must be met and not all expectations are helpful in our current context.

REFLECTION EXERCISE:

☐ **What expectations do you have of yourself that were once helpful but are no longer relevant or realistic?**

Mastering the **art of managing expectations** begins with self-awareness and reflection. As you explore the expectations you hold for yourself, you'll start to see patterns that empower your growth and others that hinder your progress. Over time, refining and realigning these expectations will help you develop a more compassionate and balanced approach to leadership and personal growth.

2. Expectations We Have of Others

We all hold expectations of other people, ranging from the simple and straight-forward to the subtle, and sometimes even surprising. These expectations influence how we see others and how we feel about them. When our expectations of others aren't met, it can feel like **broken trust**, which has significant effects on relationships and well-being—both in the workplace and beyond.

In this section, we'll explore how our expectations of others manifest and the impact they have on our interactions.

i. Expectations of Upholding My Ideals / Values

Expecting others to uphold or align with our ideals is something that occurs across a broad spectrum of life. For example, if we are devoted to a particular artist or public figure, we might admire their work, follow their social media, or even aspire to their way of life or values. We may also project our own values onto them, making the assumption that if we can relate to their output, then they must be like us in other ways too. When they change course, it can feel like a betrayal of what we expect from them, unless our rate of change is aligned.

To illustrate this, consider global pop phenomenon Taylor Swift. Her fans admire her music and her personal qualities. Now, imagine if Taylor Swift were to suddenly change her musical style and experiment with death metal. For some, this might be seen as a bold, authentic move in her artistic journey. However, for thousands of others, this shift could lead to feelings of grief and betrayal. These fans have a strong expectation of Taylor Swift remaining the clean-cut, family-friendly pop star they've supported throughout her career.

Similarly, in romantic relationships, we often idealise our partners, cherry-picking qualities we love and expect them to sustain those traits forever. This creates a heavy burden on the other person, particularly when they are growing and changing in ways we didn't expect or plan for.

This illustrates how expectations of others can be rigid and unrealistic, often leading to disappointment when they are unmet.

REFLECTION EXERCISE:

- ☐ **Has there been a time when you expected someone in your life to live up to unsustainable ideals?**
- ☐ **How do unrealistic ideals impact relationships?**

ii. Expectations in Conversations

One common expectation is the assumption that others know what we want from a conversation—this is the **'read my mind'** expectation. However, just because we know what we want, doesn't mean that anyone else does. Imagine how much more complex it becomes when we haven't fully identified what we want, AND we combine that with an expectation that someone else will be able to work it out! Miscommunication often happens because we fail to identify and/or express our expectations fully. A classic example of this occurs when two people engage in a conversation with different objectives but fail to communicate them clearly at the outset.

CASE STUDY: What is the Purpose of Our Talk?

Elaine is struggling with a colleague who seems resistant and antagonistic, making it difficult for her to do her work. She approaches her manager, Alex, hoping to discuss the issue. Elaine is someone who likes to talk through her emotions and verbalise her thoughts to gain a better perspective. Her expectation is that Alex will listen, offer her reassurance, and provide a safe space for her to process her emotions.

Alex, however, is a natural problem-solver. His expectation is that any conversation about work issues should lead to an action plan. As soon as Elaine begins discussing her concerns, Alex is already formulating solutions in his mind, interrupting her to offer pragmatic, solution-focused suggestions. Elaine doesn't feel heard, and Alex feels the conversation is unproductive. Both walk away frustrated.

What went wrong?

Elaine and Alex had different expectations for the conversation. Elaine wanted reassurance and time to process her thoughts and emotions, while Alex wanted to fix the problem and move forward. Neither approach is better or worse, but their misaligned expectations led to a disconnect.

If Elaine had been clear from the start that she needed a safe space to vent, Alex might have responded differently. Conversely, Alex could have asked Elaine if she was looking for solutions or simply wanted to be heard.

Both Alex and Elaine could have set expectations for the conversation, reducing frustration and ensuring that their objectives were aligned.

A well known technique that helps adults and teachers understand what a child wants is to ask them if they want to be 'helped, hugged or heard'. Imagine if all conversations in a work context started with an agreement about what kind of conversation was happening - miscommunication and frustration would be significantly reduced!

While it may not be appropriate to ask a colleague if they want to be 'helped, hugged or heard' the concept is still something to bear in mind. It can be as simple as adjusting the terms and the tone to make it more suitable - "Would my help, some emotional support, or a sounding board be most helpful right now?"

Deirdre Morrison and Gabriella Benkő

REFLECTION EXERCISE:

Think of a conversation where it would have helped to set expectations beforehand.

- ☐ **How could you have done that?**
- ☐ **Are there any upcoming conversations where this might be a useful approach?**

Other Expectation-Driven Elements of Conversation

What else in our conversations might be driven by expectations? Consider humour and cultural references—how often do we assume that others share the same sense of humour or understand the references we make?

Humour can be a powerful bonding tool within a team, but it doesn't always land the way we hope. Even when delivered with kindness, jokes can miss the mark if the context isn't shared. If someone doesn't get the reference, they may feel excluded or even stupid, which can undermine their confidence or participation.

This becomes especially tricky when stress is involved. Under stress, what might usually be seen as amusing can provoke tears, frustration, or conflict. Our brain's capacity to process humour and social nuance is diminished when we're under pressure.

The key is awareness. It's not about avoiding fun or humour altogether—those are important elements for bonding and even neuroplasticity. Rather, it's about being aware of people's frames of reference and the current state they're in. If someone is

31

stressed, their brain may not have the bandwidth to process humour.

If you see that a cultural reference might not be understood or that someone is too stressed to engage with humour, adjust your approach to help them feel included.

CASE STUDY: From Classics to Confusion

Working with a team manager once, Deirdre made a reference to The Matrix, a turn of the century sci-fi movie, starring Keanu Reeves as Neo, a young computer programmer, who finds himself fighting a war that pits humans against their intelligent machine overlords. It is considered a classic.

Deirdre described having knowledge of how the human brain works as being similar to seeing the 'matrix' of reality. Her client nodded in understanding, appreciating the comparison, but then remarked, "You realise I'm probably the youngest person you'll be able to use that analogy with? People younger than me won't get it." He was right.

This reminds us of how generational divides in cultural references can create communication gaps. But the analogy held true: understanding neuroscience gives us a level of insight that is even better than being able to 'download' Kung Fu skills directly to your brain.

Mastering conversations and expectations involves not only recognising our own assumptions but also being attuned to the other person's needs, frames of reference, and stress levels.

This awareness enables us to adapt our approach, ensuring clearer communication and more meaningful interactions.

Combat, Compete, or Collaborate?
Conversations can unfold with various types of energy, and expectations shape how we approach them. Sometimes, we enter conversations in a combative or competitive state of mind, expecting to score points rather than genuinely engage. This state of mind tends to foster poor listening, heightened defensiveness and increased agression.

It can also marginalise less dominant voices, silencing equally competent

participants who may feel threatened and therefore unable to contribute effectively.

Consider a meeting where certain behaviours signal competitive or controlling expectations:

- ☐ Insisting on having the last word.
- ☐ Preparing a response while someone else is speaking.
- ☐ Adopting a 'steamroller' approach, disregarding or dismissing others' inputs.
- ☐ Hijacking the conversation, steering it toward personal goals.
- ☐ Failing to provide essential information to the group.
- ☐ Making snide or passive-aggressive remarks.

When we expect a conversation to be a win-or-lose scenario, we undermine the potential for true collaboration. This way of thinking stifles opportunities for co-creation—the real treasure in any conversation. Co-creation, as we'll see, is where leadership truly shines and long-term mastery begins.

The Four Outcomes of a Decision-Making Conversation

Social change advocate Mary Parker Follett proposed that there are four possible outcomes of a meeting or decision-making conversation. Of these four, three are problematic, and only one leads to sustainable success.

First Bad Outcome: Acquiescence

In this scenario, you are effectively disenfranchised—your voice is not heard, your input is rejected or ignored, and you have no stake in the solution. You've 'rolled over' and let others make the decision. This is not an outcome that fosters growth, collaboration, or mastery. Without active participation, team members can feel disengaged and uninspired.

Second Bad Outcome: Victory

On the surface, victory may sound like a positive outcome. But if one person dominates the conversation or insists on doing things their way without considering the group, this victory becomes hollow. Without active engagement from others, success will likely be short-lived and bring with it damaging consequences that erode trust and collaboration.

Third Bad Outcome: Compromise

Compromise is often celebrated as a good outcome, but Parker Follett argued that it still falls short of true collaboration. In compromise, no one gets exactly what they want. The result is often a mix of resentment, frustration, and lack of full commitment. No one is fully satisfied, and the best solutions remain unexplored.

The Good Outcome: Co-Creation

The only sustainable path, according to Parker Follett, is co-creation. In this approach, everyone's input is valued, and the process fosters trust, strong teams, and shared commitment. Co-creation is marked by active listening, openness, and curiosity—essential hallmarks of mastery. The group doesn't just settle for compromise or bow to domination; they build something together that reflects the collective vision and goals of all involved.

As Matthew Barzun poetically described it in his book 'The Power of Giving Away Power,', co-creation transforms a group into a constellation, where each star shines brighter as part of a greater whole. This is the ultimate expression of collaborative leadership, and mastering it requires consistent reflection and adaptation.

REFLECTION EXERCISE:

- ☐ **Which of these meeting styles have you experienced?**
- ☐ **Which do you expect to encounter in your current or future leadership roles?**
- ☐ **How could you encourage more co-creativity and collaboration in your meetings?**

3 - Expectations of Control Dynamics

Control in leadership is an interesting dance, one often framed within a hierarchical structure. For emerging leaders, transitioning into a leadership role can feel like navigating a minefield of expectations—both self-imposed and from others. Where does control lie? How is it used? Do those who have control recognise their impact?

CASE STUDY: The Promotion

David was recently promoted to team leader. Before his promotion, David had naturally and authentically built strong connections with his team. He was known for his supportive, friendly manner, and he consistently gained the trust and respect of his colleagues. David was excited to step into this new role and eager to embrace the added responsibility. However, the reality of becoming a leader brought challenges he hadn't fully anticipated.

One of the first things David noticed was a subtle shift in dynamics. Colleagues who once treated him as a peer now good-naturedly referred to him as the "big boss." But David sensed something beneath the surface. Was this camaraderie or an attempt to embarrass him? Did they still see him as one of their own, or had the promotion created a division?

David felt conflicted. Was this a reflection of their feelings towards his promotion or merely their adjustment to the new team structure? Moreover, with the promotion came a shift in responsibility. Along with managing his workload, David now had to make decisions that impacted others' workloads. Team members

3 - Expectations of Control Dynamics

Control in leadership is an interesting dance, one often framed within a hierarchical structure. For emerging leaders, transitioning into a leadership role can feel like navigating a minefield of expectations—both self-imposed and from others. Where does control lie? How is it used? Do those who have control recognise their impact?

CASE STUDY: The Promotion

David was recently promoted to team leader. Before his promotion, David had naturally and authentically built strong connections with his team. He was known for his supportive, friendly manner, and he consistently gained the trust and respect of his colleagues. David was excited to step into this new role and eager to embrace the added responsibility. However, the reality of becoming a leader brought challenges he hadn't fully anticipated.

One of the first things David noticed was a subtle shift in dynamics. Colleagues who once treated him as a peer now good-naturedly referred to him as the "big boss." But David sensed something beneath the surface. Was this camaraderie or an attempt to embarrass him? Did they still see him as one of their own, or had the promotion created a division?

David felt conflicted. Was this a reflection of their feelings towards his promotion or merely their adjustment to the new team structure? Moreover, with the promotion came a shift in responsibility. Along with managing his workload, David now had to make decisions that impacted others' workloads.

Team members sought his approval for decisions, leave applications, and project assignments. Each and every one of these new responsibilities came with certain expectations.

Suddenly, David found himself grappling with the new reality of making decisions for and about others—decisions he would have previously expected to make only for himself.

David's broader view of leadership had been shaped by various experiences throughout his life. He believed, as many do, that a leader must be in control—making decisions often without extensive discussion. His expectations of himself as a leader were moulded by his former bosses, teachers, and even his parents. And as David transitioned into his leadership role, he anticipated that his colleagues would interact with him as he had once interacted with leaders in his life. But reality didn't always match his expectations.

Emerging into leadership brings tremendous opportunities, but it also brings risks and responsibilities. The way David's team now saw him, the expectations placed upon him, and his own evolving expectations of control were all converging. Mastering this transition would require David to recalibrate his approach, recognising that control is not simply about decision-making; it's about balancing responsibility, trust, and collaboration with excellent self-awareness and self-regulation.

What David needed, as do many emerging leaders, was the awareness to understand how others perceive control and leadership—and to develop a more inclusive, flexible approach that met both his needs and those of his team. Understanding that his brain was likely playing tricks on him—turning uncertainty into perceived threat—was crucial to helping David maintain perspective and avoid reactionary behaviours.

David's next step on his leadership journey involved embracing a more open, adaptive cognitive toolkit - new ways of thinking.

He had to allow room for the diverse expectations of his team, learning to navigate and co-create decisions rather than impose them. Through applied neuroscience, he would begin to identify where his brain's stress responses could hinder his ability to lead thoughtfully and effectively. This awareness would be a foundational step towards mastering leadership in an ever-evolving landscape.

REFLECTION:

- ☐ **What are your expectations of control as a leader?**
- ☐ **Are traditional, hierarchical structures effective in today's world?**
- ☐ **If you are a decision-maker, how do you ensure your decisions are made inclusively?**
- ☐ **How do you balance control with respect, ensuring you are trusted without needing to dominate?**

3. Expectations Others Have of Us

This third type of expectation extends far beyond the workplace and can bring to the surface some of our most deeply rooted beliefs and emotions.

Take your time with this section and recognise that confronting long-held expectations—whether from family, friends, or colleagues—can be a deeply emotional process. It can be helpful to find a neutral and supportive individual to act as a sounding board to explore this reflection.

The other people in our lives —family, friends, colleagues, and community members—all hold expectations of us. These expectations can be explicitly stated or unexpressed, but both can wield significant influence.

In this section, we will explore how the expectations of others impact our aspirations, relationships, careers, and overall quality of life.

i - Expectations from Family and Friends

Our families and friends are typically the people who love and support us throughout our lives. While we may feel close to some, we may distance ourselves from others. Despite this, their expectations—whether well-meaning or burdensome—can deeply affect our choices and sense of self.

Play Safe and Secure

Many family expectations stem from a place of love and a desire for security. Security is one more way of avoiding uncertainty, which you will no doubt have spotted as a sub-theme running through this book.

Take, for example, a parent who recognises potential in their child and pushes them to study hard, pursue higher education, or strive for a stable and lucrative career. While their intentions are likely well-meaning, the expectations they place on their child may be based on what they believe is best rather than the child's own dreams and desires.

Consider the generations of school students in Ireland who clamoured to get the required exam results for the university courses that were most difficult to get into. This was at least in part, because their parents' expectation was that if you got into one of these courses, then that would lead to a good, secure job. Places on courses like radiology and actuarial accounting were enormously competitive, which in turn, pushed the entry requirements even higher. Quite often, the programmes were a terrible match for the students who succeeded in gaining entry, and the expectations of career stability and reward often did not materialise.

Be Like Us

Beyond education, families may place expectations on other personal choices—where we live, whom we marry, and if or when we have children. These expectations can be especially strong when religious, cultural, or familial traditions are involved, creating pressure to follow in the footsteps of those who came before us.

Inheritance

In some families, the expectation of inheritance becomes a heavy burden. This could involve carrying on a family business or taking responsibility for inherited assets, such as land or property. Even if the asset doesn't hold any personal appeal, family pressure can be immense. People often feel compelled to follow a path that doesn't suit their talents or passions, leading to family conflict or deep personal dissatisfaction.

Socialising

Social events within a family can also be fraught with expectations. Family members may expect us to attend events, host gatherings, or include certain people in our lives and celebrations - regardless of our personal feelings. From Sunday lunches to holiday celebrations, these unspoken social expectations can place strain on relationships and personal well-being.

Make Us 'Look Good'

A more subtle, and sometimes unhelpful expectation is when parents measure their own success based on their children's achievements. In strongly catholic countries, for instance, it was often the case that families would encourage a son or daughter to take holy orders, because that reflected well on the family. Certain families are happy when their offspring have prestigious careers, or live in affluent houses. It is as though the halo effect kicks in - the parent's job must have been well done if the child exhibits commonly accepted signs of success.

Expectations of a similar nature are commonly seen with parents of kids who are competing in various activities, from sports, to academics, to beauty pageants. How much of this is what the child truly wants or needs, and how much is more aligned with the expectations of the parent is often something that bears reflection. Parents whose unnecessary or unrealistic expectations of their children are not met can sometimes act in ways that are very damaging to the developing child.

Be Grateful

Parents may also expect gratitude for their sacrifices, such as working long hours to provide for the family. While these sacrifices are significant, the expectation that a young child should understand or appreciate them can be unrealistic. Young children prioritise time spent together over financial concerns, highlighting the mismatch between what parents may expect and what children value at different stages of life.

Live the Life I Couldn't

Many parents feel they missed out on opportunities in their own lives and may push their children to pursue the dreams they couldn't achieve. While this comes from a place of wanting the best, it can sometimes result in children feeling pushed into paths they don't want to take.

Parents who feel they have missed out on something might push children in their studies, so that they will have better opportunities than they did. Or alternatively, they may feel the need to provide their children with material things, such as toys and expensive clothes, that were beyond reach in their own childhoods.

Don't Change

Family expectations may also centre around keeping things the same. Families who see a loved one changing careers or lifestyles may experience fear or resistance. This might manifest as concerns about being 'left behind' or disapproval of decisions perceived as risky. For example, leaving an established career to start a business can prompt fears and resistance within the family.

CASE STUDY: Maria's Journey

Maria, like many women of her generation, chose to become a stay-at-home parent while her children were young. By the time she was ready to return to work, her previous career had progressed so much that there was little hope of re-entering at the level she once held. On top of this, her parents were ageing and required more of her help.

Maria realised that to accommodate both her family responsibilities and her professional aspirations, starting her own business would be the best option. She took the leap and enrolled in courses to become an interior designer. She joined business support networks, built her skills, and made time in her schedule to plan her new career— trading Netflix for self-study in the evenings.

However, her family took notice. Rather than offering support, they teased her about her 'quest for world domination,' resisting the changes that were taking her away from her traditional role of being available 24/7.

To begin with, these passive-aggressive jibes hurt Maria deeply. Fortunately, she invested in her personal growth and development, which included learning how expectations could influence her decisions. Through neuroscience insights and the support of new networks, she was able to navigate her family's resistance and build a business that suited both her talents and her lifestyle.

Friends

Friends can have expectations just as impactful as family members. Depending on how and when those friendships were formed, they can influence our behaviours and choices. Are we trying to impress or compete with our friends? Are we showing care or concern? Do we feel like the friendship is balanced and equal, and nourishes us emotionally or intellectually?

Childhood and university friendships can last a lifetime or fade over time. Our expectations of friendship often evolve based on life circumstances. A university friend group that bonded over nights out may struggle to stay connected as families and work demands change the social dynamic.

Success can also complicate friendships. For example, a well-paid friend who suggests a weekend getaway may create unintentional pressure on others to join—exceeding their financial comfort zone. Expectations of loyalty and participation can lead to feelings of guilt or resentment if unmet, as both parties are navigating different realities.

Final Thought on Expectations from Family and Friends

The examples shared here are just a fraction of the countless ways in which family and friend expectations shape us. Sometimes these expectations support our growth; other times, they hold us back.

By reflecting on these expectations and recognising their influence on our careers, relationships, and self-perception, we can better understand which expectations are helpful and which need to be re-examined.

REFLECTION:

- ☐ **What expectations have you experienced from family and friends?**
- ☐ **How have these expectations shaped your career, life choices, or relationships?**
- ☐ **Which of these expectations feel helpful or unhelpful to you now?**

Expectations from Colleagues

The expectations our colleagues have of us are wide-ranging— whether they are senior, junior, from other teams, or based in different locations. We could fill a library with the various assumptions and expectations at play. However, for emerging leaders, these are the expectations that often have the most direct impact on our aspirations and growth.

Let's take a moment to reflect on David's situation, the well-respected team member who was promoted. His rise to leadership opened many doors but also came with its share of challenges.

The Autonomy / Instruction Balancing Act

One of the key challenges for any leader is finding the right balance between giving autonomy and providing enough information and instruction. Autonomy is a critical component of our sense of engagement and wellbeing. Models such as SCARF, developed by Dr. David Rock, and Be SAFE and Certain by Lori Shook, highlight autonomy as a core element of workplace effectiveness. Shook's model specifically highlights expectations, making it a particularly powerful tool for leadership. Both these tools are referenced in the resources section.

An ongoing challenge for leaders is determining how much support or autonomy each colleague requires. For example, a new team member might appreciate some extra guidance as they get to grips with their role. However, as they grow more comfortable and familiar with their role, ongoing excessive oversight can easily slip into micromanagement, signalling a lack of trust.

Micromanagement, no matter how well-intended, can undermine both confidence and creativity in your team members. It's essential to be mindful of how your behaviour shapes their sense of autonomy. It's also important not to 'throw the baby out with the bathwater' - recognition of work well done is an important part of making colleagues feel seen, recognised and valued as part of the team. Communicating this shouldn't be confused with micromanagement, and can make the difference between a motivated colleague, and one who wants to resign.

Expectations around Asking for Support

There's another aspect to consider: even if a colleague genuinely needs support, they may not feel comfortable asking for it. Past experiences or an unsupportive work environment may lead them to believe that seeking help could make them appear incompetent or vulnerable. As a leader, it's crucial to create a safe space where asking for help is not seen as a weakness but as a strength.

An additional consideration as a leader, is that we ourselves may have come from a psychologically unsafe environment, where we did not feel comfortable asking for support, and expected negative consequences as a result. The transition to successful leadership will require us to unpick these expectations and use our experiences to inform a new environment where we can both ask for and offer support in effective ways, mitigating feelings of vulnerability.

REFLECTION:

When supporting a colleague, ask yourself:

- ☐ **Have I made it clear that all requests for help and support are welcome?**
- ☐ **How often and in what ways do I reinforce that message?**
- ☐ **Am I creating an environment where people feel comfortable coming forward?**
- ☐ **How comfortable am I in admitting that I need support in my leadership?**

Encouraging 'Curious Questions'

One of the best ways to foster an open communication environment is by encouraging the use of curious questions. A client shared a successful team practice where her team members were invited to frame their observations and insights as questions, and preface them with, "I have a curious question."

This simple shift in language sets a tone of openness and exploration, signalling that no question is wrong and that collaborative curiosity is welcomed. By cultivating an environment where curiosity, support, and autonomy thrive, you are positioning both yourself and your team to grow into more effective, engaged leaders.

REFLECTION EXERCISE:

- ☐ **Are my colleagues aware that I am here to support them?**
- ☐ **Do I reinforce this often enough?**
- ☐ **What are my expectations of their abilities to operate independently? Do those expectations match the reality of the situation?**
- ☐ **If I find myself micromanaging, what am I trying to prevent? Why is it important to prevent that? (Keep asking why!)**
- ☐ **Could 'curious questions' help foster better communication and openness in my team?**

Accommodating New Work Patterns

The world of work has undergone a radical transformation. In just a few short years, we've seen an array of options emerge—remote working, hybrid setups, flexi-time, and onsite work—all forming a seismic shift in the employment landscape.

While we won't be diving into all the complexities of leadership in this new landscape within the scope of this book, it is crucial to understand how expectations around these new working patterns influence both team members and organisations.

The simplicity of the old way of working, while often restrictive, had its advantages. Yet, it also ruled out many talented people who couldn't fit the traditional mould. The trade-off for today's more flexible options is the opportunity for greater diversity, a wider pool of talent, and more innovative work environments.

Each situation related to these new working patterns is unique. Whether you're leading a fully remote team or navigating a hybrid model, the nuanced challenges you face are likely to involve expectation management. While we can't cover all these scenarios comprehensively, it's important to note that having access to applied neuroscience coaching can help you and your team manage these transitions effectively. Some of the most useful tools that can be developed through applied neuroscience upskilling include

- ☐ Better management of emotions in the workplace
- ☐ Enhanced empathy and interpersonal communication
- ☐ Stronger decision-making skills
- ☐ Greater resilience and wellbeing

If you think this could benefit you, reach out to us. Working with us might be your best next step, and if not, we'll gladly help you find an alternative.

Emotional Availability

What does it mean to be emotionally available as a leader? Why is this increasingly expected in modern leadership?

At its core, being emotionally available means you are capable of connecting with the emotions of others, especially during difficult moments. Traditionally, hierarchical systems expected everyone to deal with the leader's emotions, but rarely was it acceptable for team members to express their own emotions to that same leader.

The modern world is shifting. Concepts like psychological safety and servant leadership have reframed how we think about emotions in the workplace. Today, leaders are expected to be emotionally available, fostering trust, team cohesion, and a space where diverse perspectives contribute to innovation and growth.

Neuroscience now looks at emotions as a set of signals from our systems that can give us a lot of information with which to inform and choose better actions, reactions and interactions. The old way of 'leave your emotions at the door' was never realistic, and in fact closed off a lot of data that would have opened doors to new ways of looking at many workplace issues.

Am I Emotionally Available?

It's natural to ask, "Am I emotionally available?" And if the answer is no, can you develop this skill?

Absolutely. Building emotional availability is a mastery process—a skill that can be developed over time. It starts with becoming more aware of how your brain works and how the brains of others function in response to emotion. Understanding how the brain and emotional ecosystem works - both your own and those of the people around you - forms the basis for emotional growth.

We're bringing this concept to your attention again because it's critical to mastering expectations. Without awareness of your own emotional state and how emotions impact your interactions, you're missing the foundation for effective leadership.

How Do You Check Emotional Availability?

You can assess your emotional availability by examining how you act, react, and interact with other people's emotions.

- ☐ Do you notice a tendency to close off emotionally when a team member expresses difficult feelings? This could look like changing the topic, or finding yourself looking for some other way to avoid the exchange.
- ☐ Are you normally emotionally available but find yourself less so when you're overwhelmed or stressed? Self-care is brain-care, and your personal ecosystem, including your ability to remain emotionally available to those you lead, will be compromised when your situation is sub-optimal.

☐ On those difficult days, when your internal resources are limited, do you feel yourself shutting down to protect your bandwidth?

Managing and processing difficult emotions takes bandwidth, and if you don't have the right tools to handle them, your body-brain matrix may automatically detach from uncomfortable emotions, leaving them unresolved.

Emotions Don't Just Go Away

There will always be emotions in the workplace—this is one expectation that will almost certainly be met. How we interpret and manage these emotions depends on the skills we've developed to enhance our awareness and understand the personal ecosystem that shapes our responses.

Not all workplace emotions are difficult. Some of them are helpful and enjoyable - such as 'collective effervescence' - that buzz that comes with a team working well together and bouncing ideas off each other, and feeling a collective drive to succeed. Difficult emotions bring unpleasant feelings, which we can attempt to avoid. However, they don't simply disappear because we ignore them. Your brain is constantly trying to communicate something through the emotions you experience. When we have the right tools to decode these signals—both within ourselves and from others—we can stay emotionally available with less drain on our physical and mental resources. This builds trust, enhances team dynamics, and empowers us to make more effective decisions.

The Power of Brain Science-Backed Tools

With a toolkit rooted in accessible neuroscience, we move beyond simply trusting the process or seeing emotions as "woolly" or "touchy-feely." Instead, we gain clarity, precision, new shared vocabulary to express the previously inexpressible, and with that, a new set of choices. Mastering emotional availability through applied neuroscience doesn't just make you a better leader—it helps you empower your team to thrive, collaborate, and innovate.

Client Expectations

n any business, alongside managing expectations of colleagues, we inevitably face the expectations of clients. These expectations can be reasonable and aligned with our service, product, and marketing—or sometimes they're not.

Let's break this down with an analogy.

Imagine you started a side hustle during lockdown, focused on your love of cheese. To pass the time, you began a small YouTube channel, reviewing different cheeses each week. To your surprise, it resonated with cheese lovers around the world, and it quickly went viral.

Every week, you showcased a new cheese, gave taste tests, and offered serving or recipe ideas. The channel thrived, and before long, you saw an opportunity to capitalise on your success. Partnering with cheese makers, you offered discount codes and special promotions. Everything was going smoothly, and soon, you started selling your own cheese, along with chutney, crackers, and cheese boards.

Your commitment to quality and customer service was evident in your reviews. People expected—and received—a premium cheese experience.

But then one day, you decided to add tractors to your online store.

Hold on a minute! What happened when you read that line? It felt like a jarring shift, didn't it? It's not what anyone would expect from a cheese business.

Alternatively, let's imagine you chose to start selling highly processed cheese slices—a significant departure from the artisan cheese journey you had been on with your customers. Imagine how that would go over with your loyal base:

- ☐ "I expected more from you—so disappointed."
- ☐ "This isn't what I expected at all."
- ☐ "This isn't even cheese—what a betrayal!"

It's like Taylor Swift releasing a Death Metal album—not what her fans signed up for!

Managing Client / Stakeholder Expectations

Your clients develop expectations based on the experiences, products, and services you've consistently offered. When you pivot or make changes, it's crucial to ensure those expectations are either maintained or effectively communicated. While businesses can successfully pivot, evolve or shift directions, understanding client expectations is key to keeping them on board for the ride.

If you offer services or consultancy, then the chances are that you will need to help your clients define or manage their own expectations from time to time. A client may come to you with unrealistic ideas of what the outcome of a working relationship might be. Deirdre experienced this first hand on many occasions while working as a communications consultant. Clients would approach her firm seeking assistance in building their visibility, messaging, or media presence. But they would also come with the expectation that media coverage would directly translate into an impact on the bottom line. Helping clients - or colleagues in other departments - to understand what is possible and realistic is also part of expectation mastery.

In the world of business, reasonable expectations are those grounded in the consistent range, quality and value you've delivered. When clients know what to expect, and have those expectations met, they learn to trust you. However, unreasonable expectations can arise from miscommunication, unclear messaging, or not setting the right boundaries upfront. This is not just an issue for client relationships, but also recruitment processes.

Managing this depends on the clarity of your communication, your ability to ask the right questions, and, of course, your individual and organisational self-awareness. The better you understand your clients' and stakeholders' expectations, the better positioned you are to meet them—or even exceed them. Just as with leadership, managing expectations is a skill that develops over time. With consistent reflection and communication, you'll be better equipped to maintain strong client relationships and navigate unexpected challenges.

Section Summary: Mastering the Definition of Expectations

In this section, we explored core sources of expectations and how they shape our personal and professional lives. By breaking down the types of expectations, we've gained a clearer understanding of how they influence our behaviours, interactions, and emotional responses—and how we can start to master them.

Key Concepts:

Sources of Expectations:

☐ **Expectations We Have of Ourselves:** Self-expectations often take the form of "I should" statements, which can serve as motivation or sources of unnecessary stress. These expectations evolve as we progress through different stages of life, and not all of them remain helpful over time. Reflecting on outdated or unproductive expectations allows us to practise self-compassion and encourage personal growth.

☐ **Expectations We Have of Others:** We tend to place expectations on others to align with our values and ideals, which can lead to disappointment or feelings of betrayal when unmet. Miscommunication frequently stems from unspoken expectations, such as assuming others will intuitively know what we want in a conversation or assignement. Clearly communicating expectations in relationships and collaboration is key to reducing frustration and building stronger, more productive relationships.

☐ **Expectations Others Have of Us:** Family, friends, and colleagues all have expectations that can influence our decisions, career paths, and overall quality of life. Family expectations, like career choices or social obligations, may create pressure but don't always align with personal aspirations. Friends, too, have evolving expectations based on shared history or success levels, and these expectations can change over time.

☐ We also looked at:

☐ **Expectations of Control and Leadership:** The transition from peer to leader introduces new expectations and control dynamics. Mastering these shifts requires emerging leaders to navigate carefully. Effective leadership strikes a balance between offering guidance and allowing autonomy, ensuring team members feel supported without being micromanaged. Emotional availability is increasingly expected from leaders, as it fosters trust, psychological safety, and team cohesion, setting the foundation for long-term success.

☐ **Managing Client Expectations:** In business, mastering client and stakeholder expectations is critical, particularly when your service or product offerings evolve. Misaligned or unmet expectations can hurt client trust and damage your brand. Maintaining consistency in client expectations helps businesses build credibility, retain loyalty, and create lasting relationships.

REFLECTION POINTS FOR MASTERY:

- ☐ What "I should" statements shape your self-expectations, and are they helpful or holding you back?
- ☐ How can you refine the way you communicate expectations in conversations, collaborations, leadership roles to foster better outcomes? What preparation would be helpful?
- ☐ How have the expectations of family, friends, or colleagues influenced your decisions, and are those expectations still relevant or helpful?
- ☐ In what ways can you ensure that the expectations you set with clients or stakeholders align with the service or product experience you consistently deliver?

PART 3
The Substance of Expectations

What's Behind Our Expectations?

Our expectations, and the responses we have to them, are driven by our brain. The human brain is an incredible organ, connecting every part of our physical bodies, with everything in our lifelong range of experiences, to help navigate us through life. Your brain is both similar to and different from every other brain that has ever existed.

No two brains are alike.

Healthy brains have similar structures and biology. But they also have unique world views and survival strategies, based on what you have learned and experienced on your way to this very point in time, where you now find yourself reading this book. Past experiences shape how we perceive the present.

Right now, there are approximately **86 billion neurons** in your brain, each of which carries signals around your brain, with **trillions of potential connections**. Knowing a little bit more about why this is important, and what it looks like in day to day life helps us to understand our relationship with expectations.

At the base of the millions of years of biological evolution that have taken place to shape this amazing piece of kit, is a simple directive: **Stay alive and safe.**

This might sound very basic for a species that has built space stations, created art, AI, split the atom, and decoded DNA.

The brain's primary function is to keep us alive.When we start to think about all of our actions, reactions and interactions through this lens, the world suddenly starts to look different. It is a profound reflection point for leaders.

In the coming pages, we're going to dig deeper into the differences between expectations, assumptions and predictions, and create a better rounded understanding of the similarities and differences between them, so that we are well equipped to identify and work with them as they arise.

Digging into Subtleties.

What's the difference between something we expect, and something we're predicting, or assuming?
In many instances this looks like we're splitting hairs, but there are important distinctions.

Back at the very start of this book, we laid out some definitions, and this is a good time to check in on those again.
An **expectation** is a strong belief that something will happen or be the case, whereas a **prediction** is a statement of something that you think will happen.

An **assumption**, on the other hand, is
a belief that something is or should be a certain way, and can exist without any evidence to support the assumption.
Assumptions are the root of many biases.

Where's the difference?

Let's look at the difference between predictions and expectations first. For the purposes of this book, the difference is that a prediction taps into things that we already know or have experienced - whether they are directly connected or not.

This might take the form of how we think and feel about engaging with a certain person, task, or event, for instance. If we've met with Joe on multiple occasions over the last three years, we've had a positive experience each time, then we predict with a fair amount of confidence that we will also have a positive experience the next time. Conversely, if we've had a consistent run of unhelpful experiences with Joe, and left each one feeling frustrated or intimidated, then our prediction about this meeting might be less optimistic. We use our general and contextual data set to predict an outcome.

The danger with prediction is that we can rely on it too much. "He's always like that," is clearly a problematic statement. Your experience of him may lead you to believe that, but it is highly unlikely that he is ALWAYS like that. Additionally, it rules out the fact that people can - and do - change their approaches, thinking, strategies, and ways of looking at the world.

EXAMPLE

Michael has been managed by Marian for 18 months. In that time, he has come to recognise that she is firm but fair. Their meetings are to the point, and rarely stray from the agenda. Michael has enough experience with Marian to predict with a fair

degree of accuracy how she will behave. If there is an occasion where Marian's behaviour veers away from what Michael has known, then his brain has a few strategies to integrate this.
He may estimate that there is something out of the ordinary with Marian, and she will resume her normal behaviour soon. Or he may - depending on the nature of the behaviour - reconsider what he thought he knew about Marian. In other words, he has a choice. He might choose to rationalise Marian's behaviour if it contradicts what he has previously experienced.

Or he might choose to consolidate a previously held view. This can look a little different. For instance if he always had a feeling that Marian was more firm than fair, and didn't really like this, then he might take Marian's off day to consolidate an underlying view that she wasn't someone he liked working with.

Expectations quite often rely on imagination. This can lean in the direction of hope, such as a hope that this new role will lead to more influence in the organisation. Alternatively, it can lean in the direction of fear or anxiety. For example, an upcoming organisational change or reshuffle might create expectations of layoffs or additional work or upheaval.

Let's think about a specific situation, through a lens of expectation and prediction. When you want to see someone's face light up when you give them that gift, or deliver that piece of news. That's an expectation. If they don't react according to what you want to happen, you may experience disappointment, or feel rejection. If you think that they will act or respond a certain way, based on your past observations of giving them a similar gift, then you are creating a prediction, based on something you know or have experienced.

Why is this important?

Predictions can be every bit as tricky as expectations to manage, and because it's often hard to tell them apart, it's important to understand predictions and their impact on our outlook.

In the situation of Michael and Marian, Michael was using his contextual knowledge of Marian to make an educated guess about how he should approach the situation. This is the brain's way of identifying good strategies to succeed in a situation. Based on what it already knows, it decides how much energy will be required to navigate a course that will ensure a successful outcome - AKA survival. This may sound dramatic, but let's strip it right back to this level for a moment.

If Michael expects a reasonable meeting, which will not put his system under undue stress, based on his previous experiences, then his brain does not engage the systems that prepare him for 'dangerous' outcomes. He goes to the meeting in a relaxed manner, and does not typically feel anxious about it.

However, if his past experience had led him to believe that this would be a difficult meeting, where he would feel threatened in some way - as though he had to stand his ground, or might lose his job, for instance, then his brain will prepare for a very different kind of encounter.

This is where knowing our bodies and the way we experience feelings, sensations and emotions, becomes very important.

EXPERIENTIAL EXAMPLE

Let's use a broad brush example.

If you can remember a time when you were very excited about something - for instance, the night before your birthday or something you were really looking forward to as a kid, you might have experienced certain things. You might have had trouble falling asleep. You might have been fidgety, or your heart might have been beating faster. But something good was going to happen, so you felt excited.

On the other hand, the night before an exam or a speech, you might also have had trouble falling asleep, felt fidgety, and noticed your heart beating faster. But because you were not looking forward to the event, then it felt like nervousness, or anxiety.

But here's the thing - notice how similar the actual physical sensations mentioned are. The only difference is in what we think is going to happen - our predictions and expectations, based on our contextual knowledge and past experience.

Anxiety and excitement are very similar in terms of how the brain produces them. The difference in how we interpret the sensations in our bodies is determined by the context in which we experience them. Are we excited? Probably, if we expect something good to happen. Are we anxious? That's more likely if we expect something that is unpleasant, or leaves us vulnerable.

REFLECTION EXERCISE:

- ☐ **Think of the sensations that accompany anxiety or excitement.**
- ☐ **How do you recognise those emotions?**
- ☐ **How do you decide that they are what you are feeling?**

Let's revisit Michael and Marian. If Michael's prediction was that it would be a difficult meeting, because that was his general experience of meetings with Marian, then there's a good chance that he would go to it feeling stressed and anxious. This would put him in a sub-optimal position to engage properly with the meeting. In addition, his manner might be hesitant, or defensive, leading Marian to question his effectiveness in her team.

At this point, Michael's prediction almost becomes self-fulfilling. He is ineffective, because he is predicting a difficult situation, which creates difficulty in being effective - a dangerous cycle.

Let's think about our evolution for a moment. In many ways, everything we do is motivated by the desire to survive, and after that, to thrive. When our system senses any danger whatsoever, it takes steps to navigate that danger. Bear in mind that everything we have ever experienced, which goes right back to our very first days, and before - including the state of our mother's health and wellbeing before we were born, has become part of our alert and response system.

Aside from the biology that is at play, there are also beliefs and expectations - conscious or unconscious - that are handed down through generations.

These may be beliefs about self-worth, whether the world is a safe or unsafe place, or biases about race and gender, for example.

The Boy Who Cried Wolf

The Boy Who Cried Wolf is a classic tale of prediction, based on past experience. In the original Greek tale by Aesop, a bored shepherd boy raises the alarm that a wolf is coming to attack the sheep. He is so amused by the villagers' panicked response, that he pulls the same trick a few more times to enjoy their reactions. However, when the wolf finally does arrive, and the boy tries to get the villagers to come and help protect the sheep, they believe that he is trying to fool them again. They ignore him. As you can imagine, it did not end well for the sheep.

Let's reimagine the story of the boy who cried wolf for a moment. If we imagine that the little boy in the story is our nervous system, who initially is doing its best to protect us (whether from boredom or something more sinister), jumping at the shadows of the unknown and raising the alarm.

We learn that there are rewards in acting this way. Raising an alarm can feel like we are taking action, or being powerful, or making things more secure by being vigilant to danger, as examples. It can also create 'rushes' of various neurotransmitters, such as dopamine or adrenaline, each of which will make us feel certain ways.

Now let's put ourselves in the position of the villagers for a moment. In a situation where we have experienced something with an outcome we don't want (racing to help someone who was playing a trick, or engaging with an individual who constantly leaves us feeling bad) then our system develops ways to engage with this. It's often based on what we already know about similar situations.

Consider colleague interactions - especially ones that are consistently difficult, for example, a person who always has a problem for every solution, or can never converse without a snide remark. Maybe you've met someone who's known for bringing down the mood, or setting the team at odds with each other.

Every time we get those signals, and build on those patterns, we are becoming more and more like the villagers, who came to believe that the boy was lying, no matter what was really going on. Eventually, we can even come to see intent that isn't there, because that's what we believe about the person.

Because that's what we expect from them.

REFLECTION EXERCISE:

☐ **Is there a colleague that you expect to behave in a certain way based on your past experience of them?**

☐ **Are there colleagues you see as positive or helpful, and negative or unhelpful, based on your experience of them?**

☐ **What would change if you set these experiences aside and approached them differently?**

The Impact of Your Energy Needs

Your brain is an incredibly complex piece of kit, and in terms of your body's energy and resources, it's expensive to run!

Your brain consumes about 20% of the energy that it takes to run your system, for a typical lifestyle.

The average brain weighs about 1.5kg. That's about 2% of an adult's weight, give or take for variations in body size and shape.

Our body weight may change throughout adulthood, but generally, our brain's weight remains relatively stable. So let's think about that energy consumption in proportion to that size difference. If the rest of your body chomped through calories in proportion to your brain then, assuming you were eating a typical, balanced diet for an adult, you could only sustain a body weight of approximately 7.5kg. That's about 16.5 pounds.

The average newborn is almost half that weight!

So how does your brain manage all this? Well, it has some very interesting ways of making sure that your resources are managed.

One of the most surprising of these is bias.

Biases are shortcuts in our thinking, also known as 'heuristics' that can cut down on expensive brain activity.
Even if we rule out the biases that are based on culture, gender, and socioeconomic factors, our brain is constantly using biases to navigate the world. Here are some examples:

Negativity Bias: a tendency to notice negative things more than positive things.
Function: Noticing the negative can keep us safe, whether this is from physical or emotional danger.

Confirmation Bias: a tendency to look for ideas and information that confirm what we already believe.

Function: Our beliefs are part of our identity - and our brain has invested a lot of energy in creating this way of making sense of the world. If we find more evidence to support our existing beliefs, then we reaffirm our belief that our view is the right one. This saves us having to question and possibly change those beliefs.

Blind Spot Bias: a tendency to believe that we don't have biases. Adam Grant referred to this as 'The I'm Not Biased' bias. We are all biased in some way, so let's just bear that in mind.

Function: This bias functions in a similar way to the confirmation bias, in that it protects our identity, and our sense of being right. If we are wrong, then we have to change something in our thinking or approach, and this, as we will see next, is something our brains would often rather avoid.

Resistance to change

Change requires neural rewiring - forming new connections between the neurons in our brain. The way we currently do something represents an investment in energy that our brain has already made, and learning to do something a new way represents a new investment. If the motivation for this change isn't strong, then it makes more sense, from an energetic cost point of view, to leave things as they are.

This can play out in our ability to address and redress our expectations too.

As well as creating new connections, if we want to change something and have it 'stick' then we need to strengthen the connections. This is done through a process called myelination, when a layer of fatty wrap forms around the axon of the connecting neurons. This speeds up the ability to transmit signals between the neurons.

We can see this in action when we practise something physical - a dance move or typing, for example. To start with, our movements are clunky, and we make lots of mistakes. As we repeat the actions, our competency grows, and we find the actions becoming smoother and more effective. The repetition of the action causes the myelin to build up, making the connection of these neurons more energy effective.

However, if we've built up connections - such as a habit that we want to change - it can feel like our brain resists the change because of the energetic cost of creating new connections, when existing connections have already been created, and - in your brain's view - are working. If it ain't broke, don't change it, might be a summary of your brain's policy on such matters.

But, our ways of thinking about things, and communicating, can also be well established connections in our brain. And this means that if we try to change them, then it may feel uncomfortable, or difficult. Your brain may resist that additional energy spend, because it doesn't really see the point. The motivation to create change needs to be strong, and we can very often talk ourselves out of doing that.

Learning and change are laden with expectations.

Resistance to change is a big challenge for leaders and organisations. However, awareness of how the brain changes, and how we can support that with simple everyday actions, can make it much easier to embrace change and make it stick.

Many organisations invest large amounts of money in training and developing their people. In many instances, this can be wasted because the missing ingredient is a basic understanding of how change happens and how we can support it for sustainability.

As individuals, we also have expectations about change - sometimes that it will be easy, and other times that it's too hard, or that we can't change. All of these expectations impact our success or otherwise.

Think of it like putting in foundations - when people are empowered to embrace change because they understand their brains better, then it's much more likely that your change programmes and personal development will yield returns.

Neuroscience-based development programmes are our speciality! Don't hesitate to get in touch if you want to help your teams overcome change resistance!

Section Summary: The Substance of Expectations

This section explored the neuroscience behind our expectations and how they shape our interactions, emotions, and decisions. It highlights the connection between our brain's evolution, energy efficiency, and the ways we interpret and react to expectations, assumptions, and predictions.

Key Concepts:

1. The Brain's Role in Expectations:

- ☐ Our brain's primary function is to keep us safe and alive. This influences how we form and respond to expectations.

- ☐ The brain processes information through 86 billion neurons, creating patterns and strategies to manage our interactions and guide us through the world.

2. Predictions vs. Expectations:

- ☐ **Expectations** are often based on imagination and can lead to emotional reactions like hope or anxiety.

- ☐ **Predictions** rely on past experience and knowledge, but both can be tricky to manage, as they can create biases or lead to self-fulfilling prophecies.

3. Impact of Predictions and Emotional Responses:

- ☐ Our brains generate similar physical responses for both excitement and anxiety, differentiating the two based on context and expectations of a positive or negative outcome.
- ☐ Unmet predictions can result in heightened stress, negatively affecting performance and decision-making.

4. Bias and Energy Efficiency:

- ☐ Cognitive biases help conserve brain energy by acting as shortcuts in decision-making. For example, **Negativity Bias** helps us detect danger, while **Confirmation Bias** reinforces existing beliefs.
- ☐ Biases allow us to process information faster, but they can also distort our perceptions, especially when dealing with assumptions and expectations.

5. Resistance to Change:

- ☐ The brain resists change because it requires new neural connections, which are energetically expensive to form and reinforce. Existing patterns, whether habits or ways of thinking, feel more comfortable because they require less energy.
- ☐ Overcoming resistance requires motivation and repetition to strengthen new neural connections, making new habits more energy-efficient over time.

REFLECTION EXERCISE:

- ☐ **How do you distinguish between excitement and anxiety in your own emotional responses?**
- ☐ **Can you identify biases that shape your expectations, both in personal and professional settings?**
- ☐ **When have your expectations, based on past experiences, created self-fulfilling prophecies? How might you change your approach?**

This section invites you to reflect on how deeply intertwined your brain's functions are with your expectations and how self-awareness and knowledge of neuroscience can empower you to manage them effectively.

PART 4
Beyond Expectations

At the start of this journey, we made it clear that the goal was natural part of life—woven into every experience, thought, and interaction. Instead, our aim was to help you break down this powerful phenomenon, so that you can understand it and make more intentional choices about how it shapes your life and leadership.

By now, you've had the opportunity to explore:

- ☐ What expectations are and how they differ from predictions, assumptions, and beliefs.
- ☐ The various kinds of expectations that arise and how they subtly influence our decisions and behaviours.
- ☐ How our brain's neuroplasticity supports not only the formation of expectations but also the ability to reshape and change them when needed.

You should also have begun to engage with the expectations in your own life—whether internal or external—and developed a clearer understanding of how they affect your day-to-day effectiveness.

In this section, we take things further by introducing tools designed to help you master the expectations that arise in your personal and professional spheres. These tools are practical applications of the insights you've gained so far, enabling you to:

- ☐ Identify the origins of your expectations.
- ☐ Create a snapshot of expectations in any given situation.
- ☐ Evaluate whether expectations are helpful or unhelpful to your goals.
- ☐ Make conscious choices to retain or release expectations.
- ☐ Reflect on how expectations influence your growth, leadership, and relationships.

The tools are arranged in a logical sequence, but you are encouraged to use them flexibly. Engage with each tool as and when it's helpful—whether individually or in combination, depending on your specific needs at any given time. Applying the tools to specific situations will be more productive than trying to map every possible expectation.

You will notice that many of the tools are based on a simple quadrant graph - all you ever need to plot them out is a piece of paper and a pen! Tools that are simple to use are more likely to get used, and be useful when we need them. Complex tools can be wonderful, but they can also be time consuming and have a steeper learning curve.

Enjoy the simplicity!

"Perfection is achieved, not when there is nothing more to add, but when there is nothing left to take away."
Antoine de Saint-Exupéry

Resources

Copies of these tools are available for download on our website. You can get your printable copies by visiting https://brainex.press/BEresources

For those who prefer working on digital platforms, you'll also find information on how to implement these tools using commonly available digital apps.

Note: These tools are intended for your personal development. If you're interested in using them in a professional capacity as a coach, trainer, or facilitator, please contact us to explore practitioner licensing options.

List of Tools:

1. **Knowns and Unknowns** – A tool to clarify what information is creating leadership expectations.

2. **Mapping Sources of Expectations** – Identify and categorise expectations.

3. **The Expectation Snapshot Generator (ESG)** – Capture factors influencing your expectations in a given scenario.

4. **'F.U.C.K. the Shoulds' Model** – A 'stacked' tool incorporating several diagnostics to help analyse and assess unnecessary "should" expectations.

5. **Expectation Effectiveness Axis** – Measure how effectively your expectations serve you in order to decide whether to retain or release them.

6.   **Pre & Post Event Expectation Diagnostic & Reflection Tool** – A tool to assess expectations before and after key events in order to measure personal development and leadership growth.

1. Knowns and Unknowns.

One of the mistakes that our brains often make is thinking that what is known to us is also known to others. This fallacy often creates miscommunication, and therefore, we consider this adaptation of the Johari Window a useful tool to include in our Expectation Management Toolkit.

This diagnostic exercise is a 'brain dump' space. It is not about directly analysing expectations, but is about identifying information gaps that can impact them. This tool should be used to assess specific situations and the relevant group of people. Trying to apply it to a global situation will render it ineffective.

This simple quadrant graph can be used as a tool for continuous personal and professional development, providing you with the insights needed to foster a transparent, communicative, and effective leadership environment.

Some quadrants may not have any available information, until after the event. This is when reflection helps us to join the dots and see where patterns emerge that we can improve on.

Deirdre Morrison and Gabriella Benkő

KNOWNS & UNKNOWNS

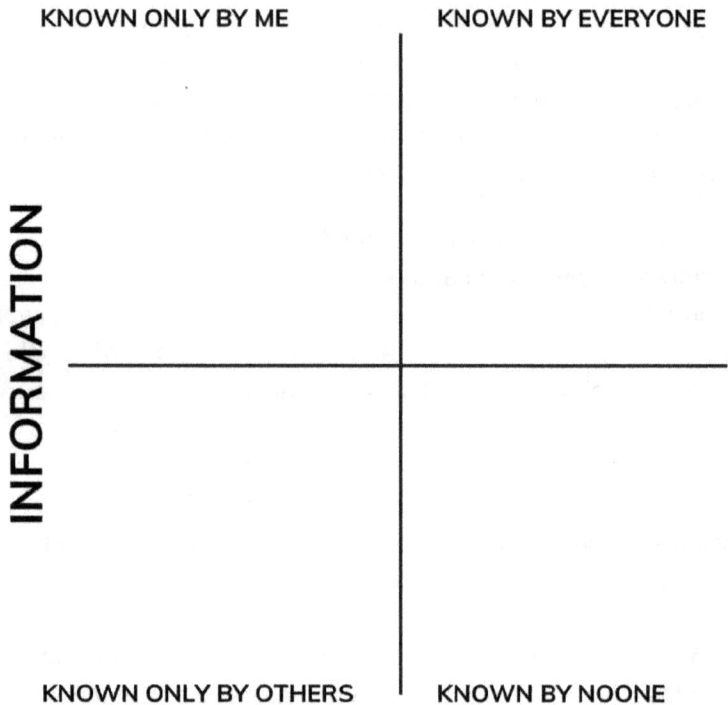

KNOWN ONLY BY ME   KNOWN BY EVERYONE

INFORMATION

KNOWN ONLY BY OTHERS   KNOWN BY NOONE

Introduction: This model focuses on information to help you navigate and clarify what is known and unknown both by you and others. This framework divides information into four quadrants:

1) Known by Me

Instruction:

- [ ] Reflect on and list the information you are aware of about the situation. This could include your personal standards, goals, or assumptions about your responsibilities.
- [ ] Consider whether some of this information would help others to manage expectations, or facilitate communication and collaboration in other ways.
- [ ] How does information, that may be known only to you, affect your daily leadership and decision making?

Why It Matters: Self-awareness in your expectations allows you to consciously align your actions with your leadership values and goals, fostering integrity and authenticity in your role.

2) Known by Everyone

Instruction:

- [ ] Identify expectations that are clear and shared among all members of your team or organisation. These could relate to performance standards, organisational goals, or operational procedures.
- [ ] Engage your team in a discussion to verify these expectations and to ensure everyone understands and agrees with them.

Why It Matters: Clarifying and affirming shared expectations prevents misunderstandings and aligns team efforts, enhancing overall effectiveness and collaboration.

3) Known Only by Others

- ☐ Seek to uncover any expectations others may have of you that you are not aware of. You can do this through feedback sessions, performance reviews, or informal conversations.
- ☐ Reflect on how these external expectations influence your interactions and leadership style.

Why It Matters: Gaining insight into what others expect from you can highlight discrepancies between self-perception and external perceptions, guiding more effective communication and relationship management.

4) Known by No-One

- ☐ Explore potential unknown expectations that might be influencing team behaviour and dynamics. These could stem from unspoken cultural norms, hidden biases, or societal pressures.
- ☐ Initiate team discussions or workshops to bring these hidden expectations to light and address their impact.

Why It Matters: Addressing and understanding these 'blind spot' expectations can significantly improve team cohesion and effectiveness by mitigating unconscious biases and promoting inclusivity. You may need to fill this quadrant in as part of a review exercise

Conclusion: This exercise is tailored to create a structured way to explore and manage the complex landscape of expectations in professional settings.

By addressing each quadrant, you canenhance not only your self-awareness but also your effectiveness as a leader. This model empowers you to navigate both spoken and unspoken expectations, leading to more informed, conscious leadership practices.

Knowns and Unknowns In Real Life.

Claire had been working with her client Michael for two years. The relationship was effective, if not as warm and personal as some of her other collaborations. From time to time, Michael was tetchy and pedantic, but overall he was fair.

One day, Claire received a call from Michael, who demanded to know why a certain decision had been taken without his knowledge, and why Claire was 'going behind his back.'

Claire was completely taken by surprise, as she had had the full support of Michael's senior team in making the decision. They were her day to day contacts, who had his authority to sign off on projects and initiatives.

Handling the call was a very intense experience for Claire, and Michael's outburst of anger made her feel very emotional during and after the call.

Let's take a look at some of the knowns and unknowns for the participants in this conversation.

Known only by Claire:

- ☐ She was doing her very best work to ensure that client objectives and deadlines were met.
- ☐ Her values guided her to honest, ethical and professional behaviour at all times.
- ☐ Her internal state during the call.

Known by Everyone:

- ☐ Michael's team managed the day to day aspects of the partnership.

Known only by Michael:

- ☐ His reasons for making the call to Claire.
- ☐ His beliefs about Claire's intentions.
- ☐ His relationship with his senior team.

Known to No One:

- ☐ Michael had an undiagnosed underlying medical condition, which caused him to act erratically.

In this example, it is clear that the unknown quadrant holds the key to this situation. However, it is also clear that Claire holds some information that she can use for her own reassurance in the moment, if nothing else. Claire knows her values, and this accusation - although unexpected and upsetting - is not based on something she has done. It's a classic case of 'it's not about me'. In this situation, Claire needs to tap into what she does know in the moment, and reflect on what she did not know later.

Michael's diagnosis helped everyone understand what had happened in this uncharacteristic situation, and he was deeply apologetic for his behaviour. Shortly after, he stepped down for health reasons.

While it may not be possible to stop and analyse a situation in detail as it's happening, asking simple questions that help us to remind us of what is known and unknown by others may well give us more effective ways to proceed.

- ☐ Is there something I would like you to know? (I agreed this with your senior execs)
- ☐ Is there something I would like to know about your perspective? (Have you talked to your senior team about this decision?)
- ☐ Is there something that we both know that we can confirm? (Is the senior team still authorised to make decisions like this for the company?)

Additionally, using this tool to reflect helps us identify which questions we might have asked, in order to identify useful ideas for future situations.

Finally, this is clearly a very unique situation, and fortunately, most interactions will not be as dramatic or have such life-changing unknowns. But each situation is unique, and should be treated as such in order to develop a finesse and flexibility in handling a variety of scenarios as they arise.

2. Mapping Expectations: A Visual Tool for Clarity and Insight

Introduction: In leadership and personal development, understanding the source and nature of our expectations can significantly influence how we interact with the world around us.

The mapping tool provided on the next page offers a visual framework to plot and categorise expectations according to their source and our
awareness of them.

This process aids in identifying dominant
trends and potential blind spots in our expectation management.

MAPPING EXPECTATIONS

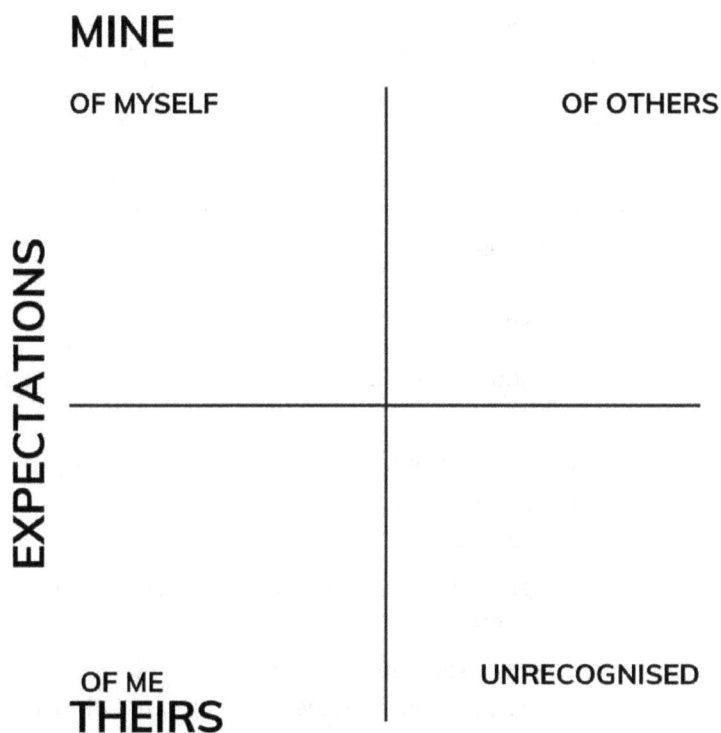

MINE

OF MYSELF   OF OTHERS

EXPECTATIONS

OF ME   UNRECOGNISED
THEIRS

How to Use the Mapping Tool

1) Identifying Expectations

☐ Begin by listing expectations related to a specific situation or general role responsibilities. These might include expectations you have of yourself, those others have of you, what you expect of others, and societal or cultural norms that influence your behaviour.

Prompt: What are you expected to accomplish? What do you expect from your team? What unspoken rules are you following?

2) Categorizing Expectations

☐ Use the grid to place each expectation into one of four categories:

☐ **Mine/Expectations of Myself:** Expectations you personally hold about your own performance and behaviour.

☐ **Mine/Expectations of Others:** Expectations you have about others' behaviour and performance.

☐ **Theirs/ Expectations of Me:** Expectations others have about you - whether you are aware of them or not.

☐ **Theirs/Cultural or Unrecognised:** Social, cultural, or organisational norms and expectations that may not even be acknowledged or discussed.

☐ **Prompt:** Consider where each expectation fits in the grid based on who holds the expectation and how visible or recognised it is.

3) Analysing the Grid

☐ Once expectations are plotted on the grid, analyse the layout:

  ☐ **Above the Horizontal Axis:** Indicates areas where expectations are easier to bring to conscious awareness, and acknowledge.

  ☐ **Below the Horizontal Axis:** Shows areas where expectations might be less clear or recognised.

☐ **Prompt:** Which quadrant is most populated? Are there areas where expectations are misaligned or conflicting?

4) Reflecting and Planning

☐ Reflect on the implications of how expectations are distributed across the grid. Consider areas requiring attention, realignment, or communication. **Prompt:** How

☐ can you address unrecognised or misaligned expectations? What steps can you take to ensure your own expectations and those of others are clear and achievable?

Conclusion: This mapping exercise provides a structured approach to visualise and manage expectations effectively. By categorising and analysing expectations in this way, you can gain insights into potential sources of conflict or misunderstanding and better align your actions with your values and goals. This tool encourages a proactive stance on expectation management, fostering a more harmonious and productive environment.

Mapping Tool In Real Life

Kai has started a new role, leading a team of 23 people. He has experience in similar roles, and has decided to analyse his expectation landscape in order to minimise miscommunication and friction as he and his team adapt to his approach.

Many of Kai's expectations are based on his past experience - as are those of his team. He starts by identifying his own expectations.

- ☐ There will be a learning curve
- ☐ He will be a good leader, and his team will trust him quickly.
- ☐ He will use his excellent bank of experience and strategic mind to make improvements
- ☐ It will be exciting and possibly a little overwhelming to start
- ☐ He could feel some doubts about his capabilities, and some 'imposter syndrome' may sneak in to his thinking
- ☐ He will be punctual, work hard, and be there for his team.

Then he listed what he expected of others

☐ That they would give him a chance and be open to new ideas he might have

☐ That they would support him in learning about the new position and their various roles

☐ That they would be professional

☐ That they would let him know if they saw anything that was going to cause problems

☐ They would like and trust him

Having looked at key expectations that he had of himself, the situation, and his new colleagues, Kai started to think about the other side of the quadrant - external expectations. He considered those that were cultural in the corporate or wider sense.

In this category, he thought about the mix of backgrounds and ages of his colleagues, how they might perceive him and his educational background. He thought about wider expectations, which are not often verbalised, but can make the difference between a comfortable and an awkward moment - like whether, in a post-Covid world, people still expected to shake hands when first meeting someone.

Finally, he considered those of his colleagues - some of which he could probably guess, but none of which he had direct information about at this stage. In this category, he decided to leave a question mark, and ask his colleagues directly about their expectations.

Looking back over the expectations that he had now filled out on the grid, and noticed how many of the expectations he placed on himself. He also noticed that he had an expectation that he would 'fix' something by bringing his experience and capabilities.

Reflecting on this, he decided to put a pin in that thought and speak to his new team first. Quite often there is a desire to 'make a mark' early on, but this isn't always the best plan. Getting to know the people, the landscape, and the systems would give Kai time to see first hand what was already effective, and where his team were open to something better.

3. Expectation Snapshot Generator

ESG - SNAPSHOT GENERATOR

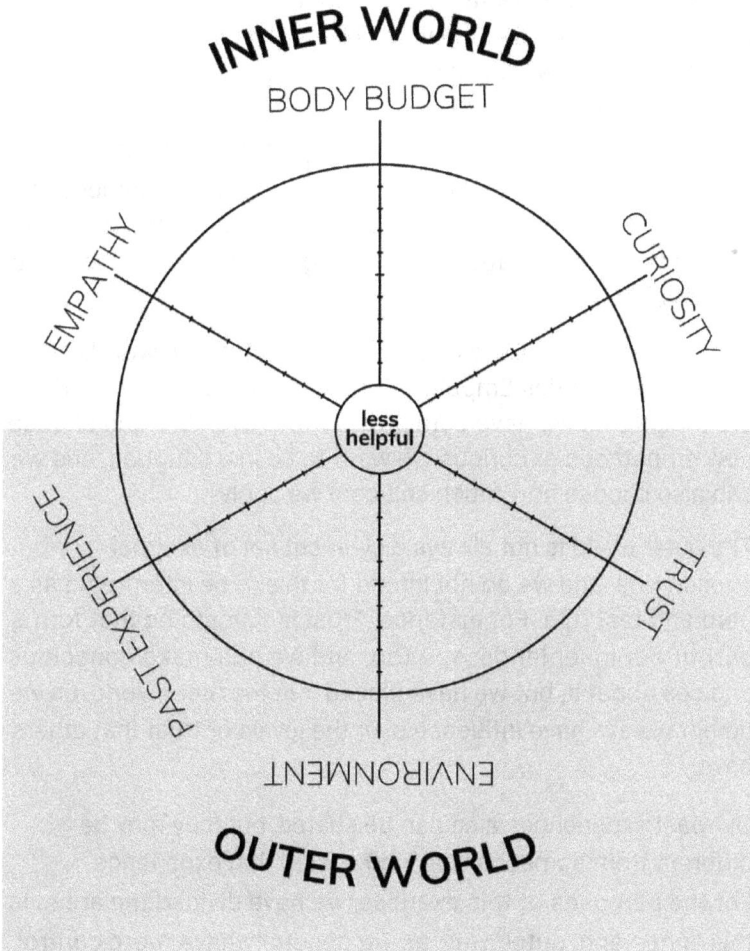

INNER WORLD

BODY BUDGET

EMPATHY

CURIOSITY

less
helpful

PAST EXPERIENCE

TRUST

ENVIRONMENT

OUTER WORLD

Expectation Snapshot Generator (ESG) Instructions

INTRODUCTION: The Expectation Snapshot Generator is a tool designed to help you quickly assess which key elements are influencing your expectations in a specific situation. Using this tool will help you reflect on whether these influences are helpful or unhelpful, allowing you to better manage your state of mind and expectations.

The ESG helps you visualise and reflect on the elements influencing your expectations. By identifying which elements are helpful or unhelpful in a given situation, you can make conscious adjustments to manage your thinking and state of mind more effectively.

The Snapshot is divided into the 'inner' and 'outer' worlds. The inner world includes Empathy, Curiosity, and Body budget. These are things that we have complete control over. We can choose how empathetic or curious we want to be in a situation, and we can also choose how much self care we apply.

The outer world is not always a clear cut set of external phenomena, and we do not intend for this to be interpreted as a hard and fast rule. For instance, Trust is something that forms part of our inner landscape too, and we can make conscious choices about it, but we have placed it in the outer world, as we do not always have influence over the levels of trust that others have.

Our past experiences also can be shared, but they may be different from someone else who shared that experience. For the purposes of this exercise, we have divided the areas in this 'inner' and 'outer' way, as we generally have more control over the inner world than the outer, and therefore can look to make changes there even if we cannot make changes to the outer world.

How to Use the ESG Wheel

1. Identify a Situation:

- ☐ Think of a specific situation where you have an expectation. This could be related to leadership, relationships, or personal goals. For example:
  - ☐ Preparing for an important meeting.
  - ☐ Managing a challenging team dynamic.
  - ☐ Setting a goal for personal development.

2. Label the Expectation:

- ☐ Briefly describe the expectation you have for this situation. For example:
  - ☐ "I expect the meeting to go smoothly."
  - ☐ "I expect my team to collaborate without conflict."

3. Assess Each Element:

- ☐ Using the wheel, assess how strongly each of the six elements is influencing your expectation. Plot each element based on how helpful or unhelpful it is to your expectation in this specific situation:
  - ☐ Helpful influence: Plot it further from the centre, toward the outer edge of the wheel.
  - ☐ Unhelpful influence: Plot it closer to the centre.

94

☐ Elements to Assess:

    ☐ Past Experience: How much are previous experiences with similar situations influencing your current expectation? Is this influence helpful or unhelpful in this context?

    ☐ Empathy: To what extent are you considering the feelings and perspectives of others involved in this situation? Is that a useful or hindering factor?

    ☐ Environment: How much does the current setting (physical or organisational) affect your expectation? Is it helping you feel prepared, or is it adding stress or overwhelm?

    ☐ Curiosity: Are you open to new outcomes or possibilities? Is this helping you stay flexible, or is a lack of curiosity causing rigidity in your expectation? Rigidity can sometimes feel like certainty that we are right.

    ☐ Trust: How much trust do you have in the people involved or in yourself to handle this situation? Is trust supporting your expectation or creating doubt?

    ☐ Embodied State: How are your physical sensations (e.g., stress, calmness, energy level) influencing your expectation? Is your physical state helping you approach the situation positively, or is it creating tension?

4. Plot Your Scores:

☐ Mark a point along each axis of the wheel to represent how helpful or unhelpful each element is:

  ☐ Closer to the centre = unhelpful or limiting.

  ☐ Further from the centre = helpful or enabling.

  ☐ Once you've plotted each point, connect them to create a shape or snapshot of your current state of expectations.

5. Reflect on the Results:

☐ After plotting the points and connecting them, take a moment to reflect on the shape of your snapshot:

  ☐ Unhelpful areas (plotted closer to the centre) may indicate factors that are holding back your ability to manage the situation effectively.

  ☐ Helpful areas (plotted further from the centre) show where positive influences are supporting your expectations.

  ☐ Consider these reflection questions:

  ☐ Which areas are pulling your expectations in a helpful direction?

  ☐ Are there any areas that are unhelpfully influencing your expectation and if so, how can you adjust those to be more helpful?

  ☐ How balanced is the overall snapshot between the Inner World and Outer World?

6. Make Adjustments:

☐ Based on the reflection, you can adjust your expectations or approach:

☐ If an element is unhelpful (closer to the centre), think about how you might reduce its influence or reframe it. For example, if Past Experience is unhelpfully shaping your expectation, try focusing on Curiosity to open yourself up to new possibilities.

☐ If an element is helpful (further from the centre), consider how you can lean on it even more. For instance, if Trust in your team is helpful, you may want to focus on collaborative approaches to reinforce that trust.

6. Revisit the Wheel:

☐ You can use the ESG wheel before, during, or after key situations to track how your expectations evolve. After the event, reflect on how the influences you plotted aligned with the actual outcome and whether the adjustments you made were helpful.

ESG in Real Life

Ali is preparing for an inter-team meeting. She expects it to be difficult because of past disagreements about how to approach a project between the teams involved.

☐ Empathy: Ali is typically an empathetic person, and can see the perspective of both teams. She places this moderately high, closer to the outer edge because understanding others' perspectives could be helpful to the outcome.

- ☐ Body Budget: Ali decides that this should be closer to outside, because she has made a point of taking a break to recharge, and made sure she has rested and had a nourishing snack for energy, and has had enough hydration.

- ☐ Curiosity: Ali places this at a relatively neutral level, as she is new to this tool, and hasn't previously thought of curiosity as an impacting factor in expectations.

- ☐ Trust: Ali rates trust low and plots it closer to the centre of the circle because she sees that she does not trust in the teams' ability to collaborate well, but also because she knows that the two teams don't have high trust in each other.

- ☐ Environment: As the meeting location is off site, and getting there requires some extra co-ordination, Ali places this closer to the centre, as she is aware it adds some stress, not just for her, but for anyone travelling to the location.

- ☐ Past Experience: Ali plots this close to the centre of the circle, because her past experience with these teams leads her to believe that the meeting will be difficult, and that is leading to anxiety.

After reviewing the emerging pattern, Ali realises that her inner world is quite well resourced, but that focusing more on Curiosity could help balance out the unhelpful influences of Past Experience and lack of Trust. By taking care to make sure her body is well looked after, she can also offset some of the stress that comes with the location issues and the inter-team dynamic.

In this instance, the only thing that Ali can't really change is the location of the meeting, but reflecting on and considering her thinking about an approach to the other influencing factors can help her to recognise her expectations and predictions, and prepare her state of mind to bring new influences to the meeting.

ALI'S SNAPSHOT

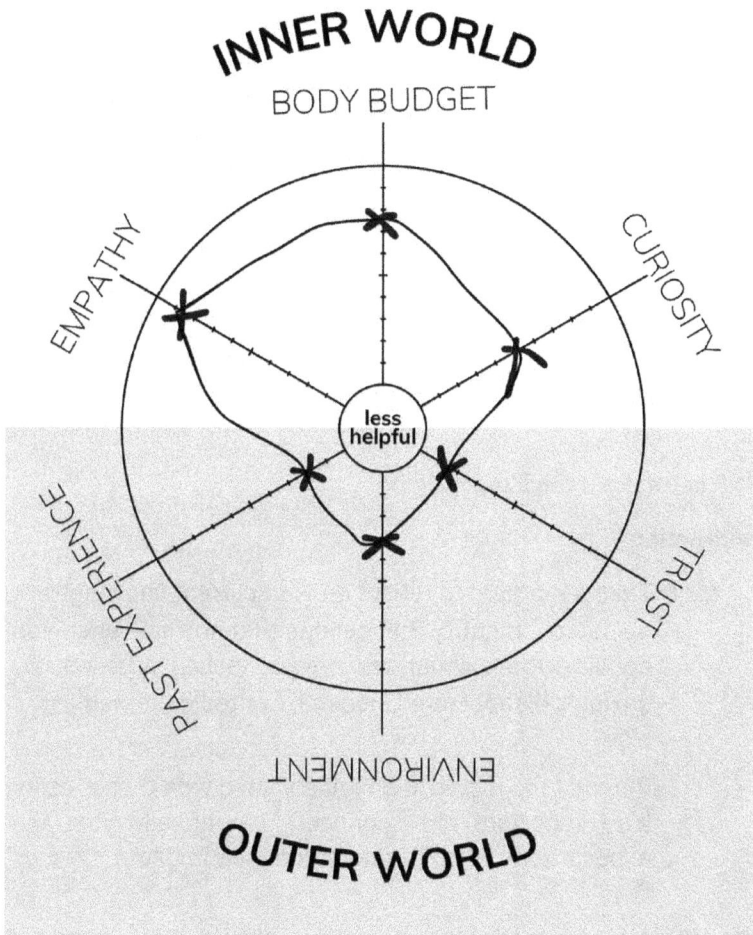

INNER WORLD

BODY BUDGET

EMPATHY

CURIOSITY

less
helpful

PAST EXPERIENCE

TRUST

ENVIRONMENT

OUTER WORLD

4. F.U.C.K The Shoulds - Managing Expectations to Enhance Leadership

Introduction: Expectations, whether our own or those imposed by others, can significantly shape our emotional landscape and influence our leadership effectiveness. The 'F.U.C.K The Shoulds' model provides a provocative, memorable, and effective framework for leaders to navigate and manage these expectations consciously. It builds on previous exercises that we have outlined, and provides a more integrated look at a situation. We would recommend using this exercise when you are more familiar with the individual exercises that we have already shared.
Here's how you can apply this model:

1) F is for: Feel the Expectations

Instruction:

- ☐ Take a moment to reflect on your current challenges or decisions. Identify the feelings that surface when you consider these situations. Are you feeling stressed, anxious, disappointed, hopeful, excited or something else?

- ☐ Pinpoint the expectations associated with these feelings. It's important to connect emotionally to these expectations to truly understand their impact.

Why It Matters: Feeling the expectations allows you to recognise the emotional weight they carry. This recognition is the first step toward managing them effectively, ensuring they do not unconsciously dictate your decisions or leadership style.

2) U is for: Understand the Expectations

Instruction:

- ☐ Ask yourself where each expectation comes from. Is it rooted in your personal values, or is it a product of external pressures from the workplace, cultural norms, or significant others?
- ☐ Analyse whether these expectations serve your purpose or hinder your progress. Understanding the source and relevance of each expectation will help you decide which ones to align with and which to adjust.

Why It Matters: Understanding your expectations in context allows you to distinguish between what is genuinely important and what may be an unnecessary burden. This clarity can enhance your decision-making and leadership authenticity.

3) C is for: Clarify the Impact

Instruction:

- ☐ Reflect on how these expectations affect your behaviour and decisions. Consider both positive and negative impacts.
- ☐ Think about the changes you would experience if these expectations were adjusted or removed. Would you be more relaxed, more focused, more innovative?

Why It Matters: Clarifying the impact of expectations helps you see their real effects on your professional performance and personal well-being. This step is crucial for rationalising which expectations are beneficial and which are detrimental.

4) K is for: Kick or Keep

Instruction:

- ☐ Decide for each expectation whether you will 'Kick' it (discard it) or 'Keep' it (maintain it). Make these decisions based on how each expectation aligns with your values and goals.

- ☐ Plan actionable steps for discarding harmful expectations and reinforcing beneficial ones. This might include setting new goals, adjusting old habits, or communicating changes to your team.

Why It Matters: This final step empowers you to take control of your expectations. By consciously choosing which expectations to keep and which to kick, you actively shape your emotional and professional landscape. We consciously choose whether to release an expectation and its impact on us, or to actively maintain its influence. This decision-making process strengthens your leadership by aligning your actions with your true objectives.

Conclusion: By implementing the 'F.U.C.K The Shoulds' model, you as a leader can transform how expectations influence your leadership. This model not only aids in personal growth but also enhances your ability to guide and inspire others. Remember, effective leadership and expectation mastery start with self-awareness and the conscious management of the expectations that shape our interactions and decisions.

5. Assessing Expectations: Keep or Kick?

Introduction:

The last step in the 'F.U.C.K the Shoulds' model encourages us to think about whether we 'keep or kick' certain expectations. But how do you decide whether something gets kicked or kept? It's not always that simple, without some analytic thinking. This tool is designed to help you make effective strategic decisions about expectations, fostering a more balanced and fulfilling approach to your responsibilities and goals.

Understanding the nature and impact of our expectations is essential in both personal growth and leadership. This tool helps categorise expectations based on their effectiveness and their level of demand (high or low), aiding in the decision-making process to either retain or release these expectations. Let's explore how to use this matrix to enhance self-awareness and decision-making skills.

Deirdre Morrison and Gabriella Benkő

'KICK OR KEEP' ASSESSMENT

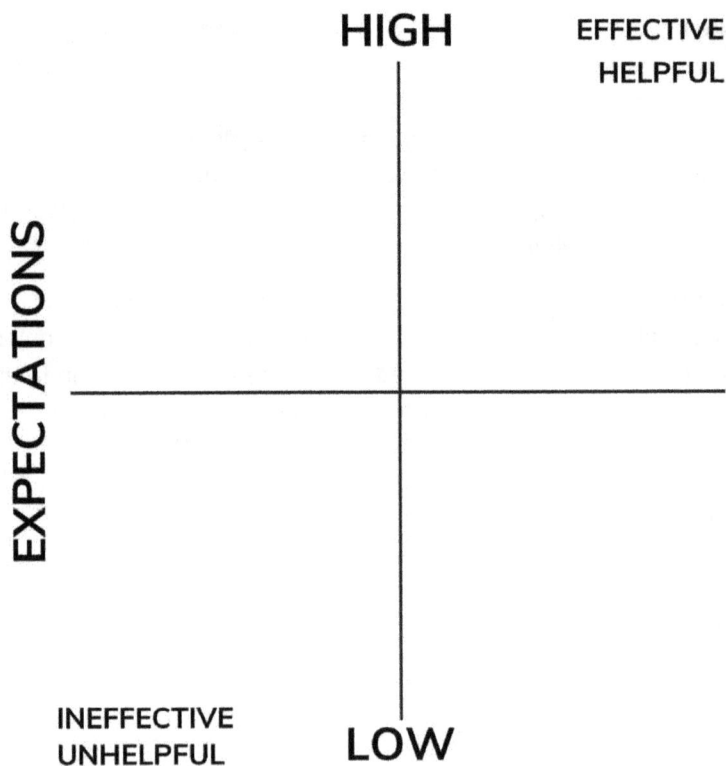

HIGH

EFFECTIVE
HELPFUL

EXPECTATIONS

INEFFECTIVE
UNHELPFUL

LOW

How to Use the Expectation Assessment Tool

1) Listing Expectations

☐ Start by listing various expectations you have or are placed upon you in relation to a particular situation you want to consider. This could include performance goals, relationship dynamics, personal standards, and more.

☐ **Prompt:** What do you expect of yourself and others? What do you believe others expect of you?

2) Placing Expectations on the Matrix

☐ For each expectation, determine how high or low the expectation is in terms of demand or standards. Then assess whether it's largely effective and helpful or ineffective and unhelpful.

☐ **High & Effective (Helpful):** These are expectations that demand a lot but significantly contribute to success and well-being.

☐ **High & Ineffective (Unhelpful):** These expectations are demanding but do not positively impact your success or well-being.

☐ **Low & Effective (Helpful):** These are minimal expectations but surprisingly effective in promoting good outcomes.

☐ **Low & Ineffective (Unhelpful):** Minimal demands and minimal benefits.

Here are some examples that will help you tease out your own specific expectations.

High & Effective (Helpful)

1. **Continuous Professional Development**: Expecting oneself to engage in regular training and upskilling. This expectation is high because it requires ongoing effort but is effective as it leads to career advancement and personal growth.

2. **Quality Standards**: Maintaining a high standard of quality in one's work outputs. This can be demanding but greatly enhances reputation, client satisfaction, and overall business success.

3. **Leadership Accountability**: Holding accountable to ethical leadership practices. It demands integrity and consistency but effectively builds trust and inspires teams.

High & Ineffective (Unhelpful)

1. **Perfectionism**: Requiring flawless results in every aspect of work. While the intention is to achieve the best outcomes, it can lead to procrastination, stress, and is often unattainable, detracting from productivity.

2. **Overcommitment**: Saying yes to every request or opportunity. This high expectation to be involved and active can lead to burnout and decreased quality of work due to limited time and resources.

3. **Absolute Control**: Expecting to control all variables in a project or team environment. While aiming for control can sometimes be beneficial, it often becomes unhelpful when it stifles team autonomy and innovation.

Low & Effective (Helpful)

1. **Delegation**: Setting an expectation to delegate minor responsibilities. It's a low personal demand but effectively enhances team capability and frees up leadership capacity for more strategic tasks.

2. **Flexibility**: Keeping expectations of rigid adherence to processes low, encouraging adaptability. This approach requires less enforcement but can lead to high effectiveness through innovative solutions and improved team morale.

3. **Regular Breaks**: Encouraging regular breaks throughout the workday. This may seem like a small expectation but can significantly boost productivity and mental health.

Low & Ineffective (Unhelpful)

1. **Minimal Feedback**: Offering minimal feedback to avoid confrontation. While it demands little effort, it is ineffective as it doesn't promote professional growth or improve outcomes.

2. **Lax Punctuality**: Having a relaxed approach to punctuality. This low expectation can undermine

professional reliability and disrupt schedules, leading to inefficiencies.

3. **Indifference to Detail**: Not focusing on details in work or communication. While this lowers the burden of scrutiny, it often leads to errors and miscommunications that could have been avoided with a little more attention.

These examples should provide a clear illustration of how different types of expectations can fit into each quadrant of your matrix, so you can assess and categorise your own expectations effectively.

3) Analysing the Matrix

☐ Look at where your expectations have clustered. Consider both the number of expectations in each quadrant and their implications. **Prompt:** Are there too

☐ many high demands that are ineffective? Are the low but effective expectations being undervalued?

4) Deciding to Keep or Kick

• Decide which expectations to retain ('Keep') and which to release ('Kick'). Use your analysis to guide this choice, focusing on reducing unhelpful stresses and enhancing productive behaviours.

☐ **Prompt:** Which expectations help you be your best self? Which ones hold you back or cause unnecessary stress?

5) Action Planning

☐ Develop a plan to address these decisions. For expectations you choose to keep, think about ways to reinforce or maximise their benefits. For those you decide to kick, consider strategies to minimise or eliminate their influence.

  ☐ **Prompt:** What specific actions will you take to implement these changes? How will you communicate your new boundaries or standards to others?

Conclusion: Using this tool to evaluate the helpfulness and demand of expectations can profoundly impact your ability to manage stress and enhance effectiveness in various areas of life. It supports a thoughtful approach to personal and professional development, empowering you to make informed choices about what expectations to embrace and which to discard.

6. Pre-Event Reflection and Planning

Classify and Reflect:

☐ **Identify a Situation:** Reflect on a specific situation where you have set expectations. This could be a project at work, a personal goal, or an upcoming event.

Define Expectations:

☐ **Highest Expectations (Best Case Scenario):** Describe what the ideal outcome looks like in this situation. What are the best possible results you hope to achieve?

☐ **Lowest Expectations (Worst Case Scenario):** Outline what the least favourable outcome might be. What are your concerns or fears regarding this situation?

Assess Current Expectations:

☐ **Current Expectations Scale:** On a scale from 1 (very low expectations) to 10 (very high expectations), rate your current level of expectation. How optimistic or pessimistic are you about the outcome?

Realism Check:

☐ **Evaluate Realism of Expectations:** Determine whether yourly current expectations are realistic, optimistic, or overly pessimistic. Reflect on the factors that influence your expectation setting.

Strategic Alignment:

☐ **Aligning Actions with Best Outcomes:** List the changes you can make or actions you can take to better align the reality with your best-case scenario. Consider adjustments in your communication style, planning methods, and mindset.

☐ **Handling Immutable Factors:** Identify aspects of the situation that cannot be changed.

Determine the most effective way to think about these factors. Are there proactive steps you can take to manage these unchangeable elements?

Post-Event Reflection

Evaluate and Learn:

- ☐ **Reflection on Outcomes:** After the event or situation has passed, reflect on how things went. Did reality align with your expectations? Why or why not?

- ☐ **Effectiveness of Alignment Strategies:** Evaluate whether the strategies you implemented to align expectations with outcomes were effective. What worked well, and what didn't?

Awareness and Management:

- ☐ **Impact of Expectation Awareness:** Reflect on how being aware of your expectations helped you manage the situation. Did this awareness lead to better decision-making or stress management?

Improvements and Insights:

- ☐ **Necessary Adjustments:** Think about what you could have done differently. How does this insight translate into a changed expectation for the future?

- ☐ **Learning for Future Situations:** Identify key learnings that you can apply to similar situations in the future. What will you do differently next time?

Identifying Blind Spots:

- ☐ **Unrealised Expectations:** Were there any expectations you had not fully realised until after the event? Where did these expectations originate, and how did they affect the outcome?
- ☐ **Unrecognised Expectations:** Explore if there were any expectations that only became apparent because they were not met. Did they influence your approach or emotional response to the situation?

The following sheet (available in the printable resources at https://brainex.press/BEresources) can help you with this exercise.

Addresssing Expectations:
Reflection, Planning & Evaluation

Make a note of your expecations for the situation in mind. Describe the best and worst case scenarios.

Rate each of your expectations, with lowest on left and highest on right. Your range of expecations will be dependent on many factors, and there may not always be a big gap between highest and lowest, or current and realistic.

Highest Expectations |ııı|ııı|ııı|ııı|ııı|ııı|ııı|ııı|ııı|ııı|

Lowest Expectations |ııı|ııı|ııı|ııı|ııı|ııı|ııı|ııı|ııı|ııı|

My Current Expectations |ııı|ııı|ııı|ııı|ııı|ııı|ııı|ııı|ııı|ııı|

Realistic Expectations |ııı|ııı|ııı|ııı|ııı|ııı|ııı|ııı|ııı|ııı|

What can you do to align outcomes more closely with best case scenario, and avoid worst case scenario? Does this influence your current expectations?

CONCLUSION
Moving Beyond Expectations

Congratulations! You've made it to the end of this book, and more importantly, you've embarked on a meaningful journey towards understanding and mastering the expectations that shape your life. Whether you've been reflecting on your own self-imposed expectations, those you have of others, or the ones placed on you, you've taken an essential step toward becoming more aware, more intentional, and more empowered as a leader.

The tools and insights you've gained are designed to be practical, flexible, and ready for real-world application. They're not about quick fixes, but about building mastery over time. Like any skill, learning to navigate and manage expectations is a process—one that involves reflection, experimentation, and growth.

What's Next?

So, where do you go from here? The answer lies in how you apply what you've learned. It's easy to finish a book and feel inspired but quickly return to old habits. We encourage you to:

- ☐ **Reflect regularly**: Continue using the tools you've been introduced to. Keep them accessible and use them whenever you're faced with significant decisions, new challenges, or shifting expectations.
- ☐ **Stay curious**: Expectation mastery isn't a one-time event; it's an active state of ongoing curiosity about

how your brain, your relationships, and your environment interact.

☐ **Be patient**: Real change takes time, and the process of managing expectations will evolve as you do. The more you practise, the more natural it will become to recognise and reframe your expectations in a way that enhances your leadership and personal effectiveness.

The Power of Support

One of the most critical insights you've likely gained is that no one achieves mastery alone. Whether it's through colleagues, mentors, or expert guidance, the journey to mastering expectations is always enhanced when we connect with others.

If you're looking for deeper insight, tailored coaching, or workshops that can further amplify your growth, we're here to help. Our work extends beyond the pages of this book, and we're passionate about continuing to support leaders like you in realising their full potential. Don't hesitate to reach out—whether for a conversation, collaboration, or personalised tools to take your leadership to the next level. We believe in the power of applied neuroscience to transform not only your approach to leadership but also the results you achieve.

Your Leadership Journey

Becoming a great leader is a continuous journey of discovery, and expectations will always be part of the landscape. But now, you're equipped to approach that landscape with greater clarity and confidence. You have the tools to identify, challenge, and change the expectations that don't serve you, and to strengthen those that do.

The path forward is yours to define—one expectation, one decision, one action at a time. With each step, you'll build not just your leadership skills but your capacity to create meaningful change in the world around you.

Thank you for trusting us to guide you on this journey. We hope you've found the insights and tools in this book to be both empowerint and practical.

And remember - mastering expectations is not just about managing the present—it's about shaping the future you want to lead.

- Stay curious.

- Keep growing.

- And always look beyond expectations.

Here's to your leadership journey!

Deirdre and Gabriella

About the Authors

Deirdre Morrison (right) and Gabriella Benkő are applied neuroscience specialists, advocating for Global Brain Literacy and empowering individuals to develop mastery in their leadership skills through enhanced brain understanding and self- awareness.

You can join them both on Linked In for more content topics like this. Follow at https://www.linkedin.com/company/brainex-press

Deirdre Morrison

Deirdre holds the Master Neuroplastician (.mNPN) designation from the Institute of Organisational Neuroscience. She has also completed extensive coach training. In her role as Chief Strategy and Innovation Officer, she has been committed to defining and formalising the competencies required for organisational neuroscience practice, bringing science-backed insights to leadership and personal development.

With a background spanning creative and communication disciplines, Deirdre has built and supported businesses for nearly three decades. Along with facilitating change, she maintains a healthy curiosity and enjoys the exploration of interdisciplinary topics. She is particularly interested in the intersection between brain science and popular culture. Her time is spent helping individuals and leaders harness the power of neuroplasticity for continuous growth. She is particularly committed to the democratisation of understanding regarding the human brain.

Deirdre is also an ambassador for the *npnHub*, a community of applied neuroscience practitioners. In her volunteer work, she provides strategic input and leadership development to community groups, helping others develop mastery in their leadership roles. She has trained in the Japanese sword art of kendo for more than a decade, with little improvement to speak of. On the plus side, this has given her much opportunity to reflect on the nature of neuroplasticity.

Visit Deirdre's Website:

https://neurocreative.studio

Connect with Deirdre on LinkedIn:

https://www.linkedin.com/in/deirdremorrison/

Gabriella Benkő

Gabriella is an ICF-certified coach who studied brain-based coaching at the NeuroLeadership Institute. As a coach, she

partners with individuals, leaders, and global teams to co-create transformative journeys toward peak performance, resilience, amplified leadership impact, and well-being. She promotes awareness of how we can expand our human potential using the latest findings from applied neuroscience.

Her lifelong learning journey, diverse interests, and curiosity began with degrees in engineering and business administration. Early in her career, she worked in finance and strategic functions at international corporations such as PepsiCo and Reynolds American Inc. Later, she created her own businesses, supporting SMEs, entrepreneurs, foreign investors, and their teams in operating sustainably and adapting to change successfully. Her passion for human development, communication, and bridging differences was at the core of her approach.

Gabriella has also promoted positive social change through her work with various non-profits, such as Ashoka and Thrive with Mentoring.

In her free time, she enjoys outdoor activities, thought-provoking conversations, and contemporary art. When asked to choose between the mountains and the beach, she opts for the peaks and valleys. As for dogs or cats, she chooses both.

Connect with Gabriella on LinkedIn:
https://www.linkedin.com/in/gabriellabenko/

Glossary of Terms

Assumptions

Beliefs accepted as true without proof or evidence. In the context of this book, assumptions often influence expectations and can lead to miscommunication or disappointment if left unchecked.

Cognitive Reappraisal

A technique used in applied neuroscience that involves changing the way one thinks about a situation to alter its emotional impact. It's a key tool in managing expectations.

Collaboration

A cooperative process where individuals work together toward a common goal. True collaboration involves active listening, openness, and co-creation.

Communication Styles

The ways in which individuals convey information, which can be influenced by cultural, personal or situational factors. Understanding and adapting to different communication styles is essential for managing expectations.

Control Dynamics

The shifting balance of authority and autonomy, particularly relevant in leadership roles. Control dynamics often play a significant role in how expectations are managed between leaders and their teams.

Cultural Expectations

Norms and behaviours that are socially accepted or expected within a particular culture. These expectations can significantly influence personal and professional relationships, often unconsciously.

Emotional Availability

The ability of a leader to connect with and acknowledge the emotions of others, fostering psychological safety and trust. Emotional availability is crucial for building strong teams and maintaining a positive work environment.

Expectations

Strong beliefs that something will happen or should happen. Expectations shape behaviour, emotions, and decision-making, and they can be internal (self-imposed) or external (imposed by others), consciously recognised, or exist beneath conscious awareness.

Expectation Effectiveness Axis

A tool to evaluate expectations as either helpful or unhelpful, guiding decisions about whether to retain or release them based on their impact on personal and professional goals.

Expectation Snapshot Generator (ESG)

A tool designed to create a clear picture of your expectations for a specific situation or event. This helps in aligning reality with expectations and identifying potential gaps.

Inner Dialogue

The conversation or voice that we hear in our head. This is referred to by various names, including internal conversation, inner critic, inner monologue, internal chatter.

Johari Window

A psychological tool used for understanding self-awareness and interpersonal relationships. In this book, the Johari Window model is adapted to help assess "knowns and unknowns" related to expectations.

Managing Expectations

The process of recognising, communicating, and aligning expectations in both personal and professional contexts to reduce conflict and increase satisfaction.

Neuroplasticity

The brain's ability to reorganise itself by forming new neural connections throughout life. Neuroplasticity plays a central role in how we adopt, change, and manage expectations.

Predictions

Statements or beliefs about what will happen in the future. Unlike expectations, predictions are based on logic or evidence rather than emotional or subjective factors.

Psychological Safety

A work environment where individuals feel safe to express themselves without fear of negative consequences. Psychological safety is foundational for innovation, collaboration, and effective leadership.

Reflection

A key practice in self-awareness, reflection involves looking back at one's actions, emotions, and thoughts to gain insight and make informed decisions about future behaviour. It is frequently used throughout the book as a tool for managing expectations.

SCARF Model

A framework developed by Dr. David Rock that highlights five factors affecting human social behaviour: Status, Certainty, Autonomy, Relatedness, and Fairness. These factors are closely tied to expectations and can be used to guide leadership and collaboration.

Self-Expectations

The expectations we place on ourselves, often in the form of "I should" statements. Self-expectations can be helpful or unhelpful, depending on how they align with our values and current circumstances.

Servant Leadership

A leadership philosophy that prioritises the growth and wellbeing of team members over the self-interest of the leader. It is closely related to emotional availability and managing expectations with empathy.

Shoulds

A shorthand description of common internal expectations that take the form of "I should" or "They should," often leading to stress or frustration. Identifying and interrogating "shoulds" is a key part of managing expectations.

Threat Response

A survival mechanism in the brain that activates emotional and physiological responses when expectations are unmet, often leading to fight, flight, or freeze reactions. Understanding this response helps leaders manage stress and make more effective decisions.

Tools for Managing Expectations

A range of practical exercises and models introduced in the book to help identify, interrogate, and manage expectations. These tools provide hands-on ways to enhance awareness and make strategic choices

Additional Resources

Here is a selection of resources to help you deepen your
These tools and readings will guide you as you continue your
journey towards mastering expectations and enhancing your
leadership capabilities.

Cognitive Biases

- [ ] **50 Cognitive Biases You Should Know**
- [ ] www.titlemax.com/discovery-center/50-cognitive-
  biases
- [ ] A detailed breakdown of 50 common cognitive biases
  that shape how we make decisions and perceive the
  world.
- [ ] **Visual Capitalist - Every Single Cognitive Bias**
- [ ] www.visualcapitalist.com/cognitive-bias
- [ ] A visual guide to every cognitive bias you might
  encounter.

SCARF and Be SAFE and Certain Models

- [ ] **David Rock's SCARF Model**
- [ ] www.mindtools.com/scarf-model
- [ ] Explore how the SCARF model can help leaders
  understand social cues that influence team
  collaboration and performance.

☐ **Be SAFE and Certain Model**

☐ www.shooksvensen.com/be-safe-model

☐ Learn how this model highlights the critical expectations in leadership interactions.

Emotional Availability and Leadership

☐ **Building Emotionally Available Leaders**

☐ www.forbes.com/emotionally-available-leaders

☐ Understand the importance of emotional availability in leadership and how it drives better team performance.

☐ **Self-Awareness as a Leadership Tool**

☐ www.hbr.org/self-awareness-leadership

☐ Discover how self-awareness can contribute to more effective leadership.

Neuroscience and Neuroplasticity

☐ **Neuroplasticity Explained in 2 Minutes**

☐ www.youtube.com/neuroplasticity

☐ A quick, visual introduction to neuroplasticity and its impact on learning and leadership.

Acknowledgements

An incredible number of people have contributed to this book in terms of support, inspiration, and sustenance.

Our collective thanks goes to Sinéad and Daisy, who opened their doors to us, and plied us with scones as we embarked on this journey. Our thanks also to Ellen, who made the trek with us, inspiring us with her energy and exercise regime. And of course Jisoo, who sent us the best care parcel ever, ensuring that Irish cuisine took a back seat for the whole week while we dined on the finest Korean delicacies.

To our coachees, thank you for bringing the topic of expectations to our sessions and inspiring us to explore them more deeply. Your challenges and insights have motivated us to find new perspectives and share them through this book.

A special thanks to Sharon Moore and Daphne Chisholm-Elie for your unwavering support, your ideas on leadership and agile personal development, and your patience in listening to our neuroscientific musings. Deeply grateful to Des Golden, a true partner in crime in spreading the word about the neuroscience of peak mental performance—our journey together makes this work all the more meaningful. Finally, Ji Soo Yoo, your support, philosophical questions, and insatiable curiosity have opened new doors for us to explore.

Thanks also to Lori Shook, who sparked an interest in the topic of expectations many moons ago, and Margaret Hayes, who has been a constant companion and source of such wit and warmth along the way. Thank you to Svetlana Bogdanova for sharing

Deirdre Morrison and Gabriella Benkő

'curious questions' as a framing device. Colleagues at ION, the Institute of Organisational Neuroscience, deserve a mention for both feeding and fanning an ongoing interest in neuroscience - Professor Justin Kennedy, Gordana Kennedy, Liz Guthridge, Greg Pitcher, John Wightkin, Shirene Dovey, and the rest of this amazing community - you are making the world a better place with your dedication to sharing and integrating this cutting edge knowledge. Eternally grateful to John Kelso for his wisdom, patience, humour, encouragement and appreciation of the entirely random.

Speaking of random, it is important to acknowledge the incredible good fortune that we have had in meeting each other. In one of the internet's finer moments, we were assigned to a zoom room during a wonderful event hosted by Alisa Cohn and Dorie Clarke. One thing led to another, and before long, we were planning events, spreading the word about how useful brain science is, and finding out that our skill sets complemented each other very well indeed! So, thank you Alisa and Dorie!
Those who read our drafts - Kate Gormely, Nick Muldoon and Francesca Schneller - you are greatly appreciated.
And of course, to our families, who have been on this white knuckle ride with us.

Thank you for expecting us to succeed.

www.ingramcontent.com/pod-product-compliance
Lightning Source LLC
Chambersburg PA
CBHW070933210326
41520CB00021B/6922